Food Industry Briefing Series

WINE PRODUCTION: VINE TO BOTTLE

Keith Grainger

and

Hazel Tattersall

Blackwell Publishing Ltd
Editorial Offices:
Blackwell Publishing Ltd, 9600 Garsington Road, Oxford OX4 2DQ, UK
Tel: +44 (0)1865 776868
Blackwell Publishing Professional, 2121 State Avenue, Ames, Iowa 50014-8300, USA
Tel: +1 515 292 0140
Blackwell Publishing Asia Pty Ltd, 550 Swanston Street, Carlton, Victoria 3053, Australia
Tel: +61 (0)3 8359 1011

First published 2005 by Blackwell Publishing Ltd

Library of Congress Cataloging-in-Publication Data
Grainger, Keith.
Wine production : vine to bottle/Keith Grainger and Hazel Tattersall.
p. cm. — (Food industry briefing series)
Includes bibliographical references and index.
ISBN-13: 978-14051-1365-6 (pbk.: acid-free paper)
ISBN-10: 1-4051-1365-0 (pbk.: acid-free paper)
1. Wine and wine making. I. Tattersall, Hazel. II. Title. III. Series.

TP548.G683 2005
641.2′2—dc22
2005005238

ISBN-13: 978-14051-1365-6
ISBN-10: 1-4051-1365-0

A catalogue record for this title is available from the British Library

Set in 10/13pt Franklin Gothic Book
by Integra Software Services Pvt. Ltd, Pondicherry, India
Printed and bound in India
by Replika Press Pvt. Ltd

The publisher's policy is to use permanent paper from mills that operate a sustainable forestry policy, and which has been manufactured from pulp processed using acid-free and elementary chlorine-free practices. Furthermore, the publisher ensures that the text paper and cover board used have met acceptable environmental accreditation standards.

For further information on Blackwell Publishing, visit our website:
www.blackwellpublishing.com

Contents

Series Editor's Foreword

In the old wine producing countries and regions of Europe, wine has been easily available for generations and is a well established part of the dietary cultures of many nations. In more recent times the beverage has increased its visibility in the diets of people in countries not blessed with an indigenous wine industry, and the change is still underway. Now wines produced in the many, and increasing number of, wine producing regions of the world are being exported globally. Consumers from widely differing socio-economic groups are able to experience wine, and no longer can it be thought of as the preferred beverage of the privileged classes. Wine has come of age in the modern consumer marketplace. It is something that is enjoyed by all kinds of consumers, often at the expense of other beverages.

A great deal of the credit for what we might term the democratisation of wine must go to New World wine producers in countries such as Australia, New Zealand, California and South Africa, as well as some adventurous Old World producers. They have worked hard over many years to produce high quality wines that appeal to modern tastes, and have developed highly efficient and flexible distribution networks that serve a global food marketplace. This said, credit must also go to the supermarkets which have helped to develop the consumers' interest in wine, significantly by creating ways to inform people about wine. No longer does one have to be an expert to find and enjoy a drinkable wine. A great deal of useful information is now routinely provided on bottle labels and supermarket shelves, thus guiding judgement, easing decision making, and increasing the likelihood that the money spent on a bottle of wine will be rewarded by the liquid in the bottle. Some people though, wish to know more about wine than can be found on the bottle or in the supermarket, or even conveyed by the TV wine expert. They need, therefore, to delve into appropriate texts, and this book has been written for them, irrespective of whether their interest is that of the enthusiastic amateur or at a more professional level.

Wine Production: Vine to Bottle has been written by two very experienced wine educators, who have drawn on many years of practical involvement with the wine industry. They have created a work that fits well between the coffee table publications that accompany TV programmes and the specialist books written for viticulturalists and oenologists. It will be of great value to those who wish to understand what wine is and how it is produced – from vine to bottle. It will also provide a very firm foundation for the development of specialist knowledge of the topic for those who wish to make a serious study of wine. It makes an ideal starting point for a process of study which gradually takes in more advanced publications. In this respect, the authors, Keith Grainger and Hazel Tattersall, have intentionally set out to produce a readily accessible text which can be easily read and efficiently assimilated. As editor of the work, I can see they have achieved their objective, and I am sure the book will be of use to professionals and non-professionals alike. Indeed, I am sure that those with an interest in wine production as well as those who work in the wine industry will find the book of great utility. Also, I expect it to be of importance to academics and students in fields ranging from food science and technology to hospitality management and food service.

This book is the third publication in Blackwell's *Food Industry Briefing Series* and it makes a worthy contribution as it deals with a specific product type. The series was created to provide people with an interest in food production and processing, and food science and technology, with condensed works that can be read quickly. The series is of particular benefit to busy food industry professionals who do not have the time to tackle heavy texts, but who wish to gain knowledge easily on certain topics. The series will also be of benefit to academics who may use the books to support courses and modules, as well as students studying the subjects of food and drink. In this respect it has been Blackwell's intention to keep the books sensibly priced, so that students can afford to buy their own copies, and college and university libraries can extend to multiple ownership.

In an age when textbooks and journals are being priced out of the academic market, ensuring access to learning through a sane cost structure is important and the publishers must be commended for this.

Ralph Early
Series Editor, *Food Industry Briefing Series*
Harper Adams University College

Preface

There are, of course, many detailed books on the topics of grape growing and winemaking, and the multifarious individual aspects thereof. These books, although very valuable to oenology students, grape growers and winemakers, are often highly technical. There are also several 'coffee table' books, which paint a picture that will be appreciated by consumers and those with merely a passing interest. Our aim in writing this book is to provide a concise, structured yet readable, understanding of wine production, together with a basic source of reference.

Although the book contains necessary scientific information, it is designed to be easily understood by those with little scientific knowledge. As such, it may prove valuable to food and beverage industry professionals, wine trade students, wine merchants, sommeliers and restaurateurs, those entering (or thinking of entering) the highly competitive world of wine production, and all serious wine lovers.

The information contained herein is not from any parochial or polarised viewpoint, although the authors, like all wine lovers, cannot (and do not wish to) claim to be totally objective. During the research and preparation we have spent much time in wine regions in both the New World and Old. We obtained diverse and detailed viewpoints from hundreds of practitioners, including growers and vineyard workers in both cool and hot climates, family winery owners, winemakers and technicians working with large scale producers, consultants and representatives of wine institutes.

We wish to thank everybody who gave us their valuable time and were happy to share their knowledge and opinions. We also wish to thank Ralph Early, series editor for the *Food Industry Briefing* books, for his valuable suggestions and work on editing the text, and Nigel Balmforth and Laura Price of Blackwell Publishers who have given us the opportunity to produce this work.

Keith Grainger
Hazel Tattersall

Acknowledgements

We are grateful to the following people and organisations for permission to use the figures identified here:

Christopher Willsmore, Warwickshire for Figs. 1.1, 1.3, 1.4 and 4.1.

Corinne Delpy, Yaslin Bucher S.A., Chalonnes-sur-Loire, France, for Figs. 9.1 and 12.2.

Mauro Ricci, Della Toffola S.p.A., Signoressa di Trevignano, Italy, for Fig. 11.4.

Nicola Borzillo, REDA S.p.A., Isola Vicentina, Italy, for Fig. 12.5.

Lynn Murray, Champagne Taittinger & Hatch Mansfield Agencies, Ascot for Fig. 15.1.

Frédérick Questiaux, Oeno Concept S.A., Epernay, France, for Fig. 15.2.

Introduction

No other beverage is discussed, adored or criticised in the same way as wine. To a few, it is something to be selected with the greatest of care, laid down until optimum maturity, carefully prepared for serving, ritually tasted in the company of like-minded people using a structured technique and then analysed in the manner of both the forensic scientist and literary critic. To many, it is simply the bottle bought in the supermarket according to the offer of the moment, drunk and perhaps enjoyed on the same day as purchased. To those favoured with living in wine producing regions it is often the beverage purchased from the local producers' cooperative from a dispenser resembling a petrol pump, taken home in a five or ten litre containers and drunk with every meal.

There is a wonderful diversity in the styles and quality of wines produced throughout the world, promoting discussion and disagreement amongst wine lovers. The wines of individual producers, regions and countries rise and fall in popularity according to consumer and press and TV media perceptions of style, quality, fashion and value. Consumers do not remain loyal when they perceive that their needs and wants are better met elsewhere. In the 1980s, wines from Bulgaria were very popular in the United Kingdom and Australian wines almost unheard of. By 2005 the wines of Australia held the top position in the UK market, by both volume and value of sales.

Few would dispute that the standard of wines made today is higher than at any time in the 8000 years or so of vinous history. The level of knowledge of producers, and thus the ability to control the processes in wine production, could only have been dreamt of even 40 years ago. Yet when *Decanter* magazine compiled a list of the greatest wines of all time in August 2004, the top position was awarded to Château Mouton Rothschild 1945, and six of the 'top ten' wines were produced more than 40 years ago. Also, in the past few years, globalisation and consolidation of producers have perhaps had the detrimental effect of producing technically good wines whose styles have become standardised. In other words, the wonderful diversity we referred to is under threat.

This book details how wine is produced, from vine to bottle. Many of the concepts are simple to grasp, others more complex. However, we need to stress at this stage that there is no single, unquestioned approach to wine production. Many procedures in common usage remain subject to challenge. Indeed, if you talk to 50 winemakers you are likely to hear a hundred different viewpoints, and many producers are constantly experimenting and changing techniques.

In considering wine production, there are two distinct stages: the growing of grapes (**viticulture**) and turning grapes into wine (**vinification**).

Throughout the wine producing world there are many in the industry, and many businesses, that carry out just one of these stages: growers who make no wine but sell their grapes to a wine producing firm, or who are members of a cooperative that makes wine. Then there are wine producers who have no vineyards, or insufficient to supply their grape needs, who buy grapes from growers small or large.

The decisions made and operations undertaken in both the vineyard and winery will impact upon the style and quality of the finished wine. These decisions will be based on numerous factors: geographical, geological, historical, legal, financial and commercial. The resources and availability and cost of local labour will have a major impact upon the decisions made and the structure of the wine production operation. Both the grower and the winemaker are aiming for maximum control: yield, quality, style and cost. Of course, the aim is to make a profit.

Grapes contain all that is basically necessary to make wine: the pulp is rich in sugar and yeasts are contained in the bloom which looks like a white powder on the skins. However, many winemakers choose to inhibit these natural yeasts and use cultured yeasts for fermentations. It should be noted that, unlike in the production of beer (and many spirits), water is not generally used as an ingredient in winemaking. The grapes should be freshly gathered, and ideally the winemaking should take place in the district of origin. However, this is not always adhered to, particularly with regard to inexpensive wines. It is not uncommon for grapes or grape must to travel from one region to another, or sometimes even to another country, prior to fermentation.

Wine is, of course, alcoholic. The alcohol in wine is **ethanol**, otherwise known as ethyl alcohol. Although it is a natural product, ethanol is toxic, and can damage the body if taken in excess. The alcohol is obtained from the fermentation of **must** (see Box I.1), by the action of enzymes of yeast which convert the sugar in the must into ethanol and carbon dioxide.

Box I.1 Definition of must

Must: Grape juice and solids before fermentation.

Although the fermentation lies at the heart of winemaking, every other operation will impact upon the finished wine. The entire production process may take as little as a few weeks for inexpensive wines, or two years or more for some of the highest quality wines. In the case of some liqueur (fortified) wines, the production process may take over a decade.

Quote: 'If you ask me: "Have you made the best wine, or the best wine that you can?" and I answer "Yes", then you must take me away and bury me.' Beyers Truter, South African winemaker, famed for making perhaps the best Pinotage in the world.

Viticulture – The Basics

The harvesting of healthy, ripe grapes is the end of a successful annual vineyard cycle and the beginning of the work in the winery. The grower and winemaker are both aware that any deficiencies in the quality of fruit will affect not only quality but also profitability. Although the juice of the grape is seen as the essential ingredient in the winemaking process, other constituents also have roles of varying importance.

1.1 The structure of the grape berry

Fig. 1.1 shows a section through a typical grape berry.

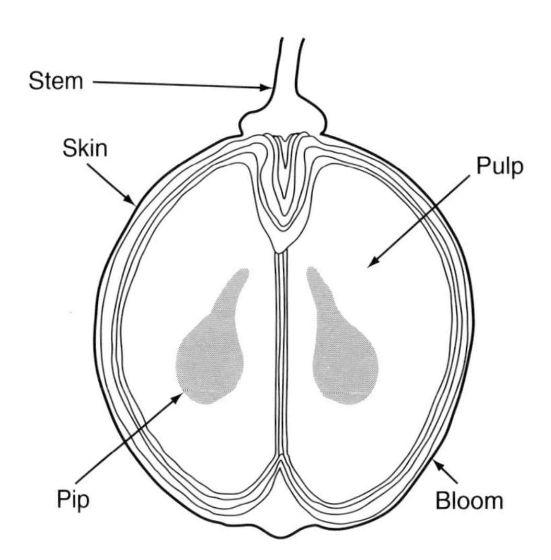

Fig. 1.1 Structure of the grape berry

1.1.1 Stalks

Stalks contain tannins that may give a bitter taste to the wine. The winemaker may choose to destem the grapes completely before they are crushed. Alternatively, the stalks, or a small proportion of them, may be left on to increase the tannin in red wine to give extra structure. However, if the stalks are not removed, they perform a useful task in the pressing operation by acting as drainage channels.

1.1.2 Skins

Skins contain colouring matters, aroma compounds, flavour constituents and tannins. The outside waxy layer with its whitish hue is called **bloom**. This contains yeasts and bacteria. Below this we find further layers containing complex substances called polyphenols, which can be divided into two groups:

(1) *Anthocyanins (black grapes)* and *flavones (white grapes)* give grapes their colour and as phenolic biflavanoid compounds they form antioxidants and thus give health-giving properties to wine.

(2) *Tannins* are bitter compounds that are also found in stalks and pips. They can, if unripe or not handled correctly, give dried mouth feel on the palate. Tannin levels are higher in red wines where more use is made of the skins and stalks in the winemaking and with greater extraction than in white and rosé wines. Some varieties such as Cabernet Sauvignon, Syrah and Nebbiolo contain high levels of tannins, others such as Gamay have much lower levels.

1.1.3 Yeasts

Yeasts are naturally occurring micro-organisms which are essential in the fermentation process. Yeasts attach themselves to the bloom on the grape skins. There are two basic groups of yeast present on the skins: wild yeasts and wine yeasts. Wild yeasts (mostly of the genus *Klöckera* and *Hanseniaspora*), need air in which to operate. Once in contact with the grape sugars, they can convert these sugars to alcohol, but only up to about 4% alcohol by volume (ABV), at which point they die. Wine yeasts, of the genus *Saccharomyces*, then take over and continue to work until either there is no more sugar left or an alcoholic strength of approximately 15% has been reached, at which point they die naturally.

1.1.4 Pulp

The pulp or flesh contains juice. If you peel the skin of either a green or black skinned grape, the colour of the flesh is generally the same. The actual juice of the grape is almost colourless, with the very rare exception of a couple of varieties that have tinted flesh. The pulp/flesh contains water, sugars, fruit acids, proteins and minerals.

- *Sugars*: when unripe, all fruits contain a high concentration of acids and low levels of sugar. As the fruit ripens and reaches maturity, so the balance changes, with sugar levels rising and acidity falling. Photosynthesis is the means by which a greater part of this change occurs. Grape sugars are mainly represented by fructose and glucose. Sucrose, although present in the leaves and phloem tubes of the vine, has no significant presence in the grape berry. As harvest nears, the producer can measure the rise in sugar levels by using a refractometer, as illustrated in Fig. 1.2.
- *Acids*: by far the most important acids found in grapes are tartaric acid and malic acid, the latter being of a higher proportion in unripe grapes. During the ripening process, tartaric acid then becomes the principal acid. Tartaric acid is not commonly

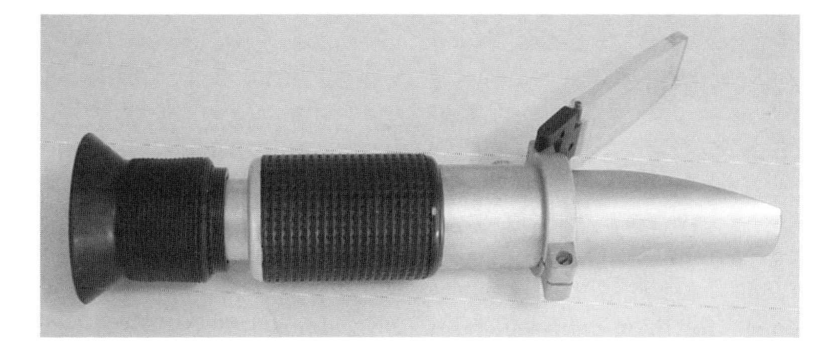

Fig. 1.2 Refractometer

found in plants other than vines. Acids have an important role in wine in giving a refreshing, mouth-watering taste and also give stability and perhaps longevity to the finished wine. There are tiny amounts of other acids present in grapes, including acetic and citric.

- *Minerals*: potassium is the main mineral present in the grape pulp, with a concentration of up to 2500 mg/l. Of the other minerals present, none has a concentration of more than 200 mg/l, but the most significant are calcium and magnesium.

1.1.5 Pips

Pips or seeds vary in size and shape according to grape variety. Unlike with stalks, there is no means of separating them on reception at the winery. If crushed, they can impart astringency to the wine due to their bitter oils and hard tannins. As we shall see later, today's modern presses take account of this.

1.2 The grape vine

The grape variety, or blend of grape varieties, from which a wine is made is a vital factor in determining the design and style of the wine. However, it is **not** the only factor, although many a Chardonnay or Cabernet Sauvignon drinker believes otherwise. Wines made from a single variety are referred to as **varietals**. The name of the variety may be stated on the label, this concept having been introduced in Alsace in the 1920s and promoted heavily by the Californian producers in the 1970s, has now become commonplace. However, many wines made from a single variety do not state the fact on the label, e.g. a bottle of Chablis AC will rarely inform you that the wine is made from Chardonnay, indeed the rules of **Appellation Contrôlée** often do not permit the name of the grape to be stated. Many top quality wines are made from a blend of two or more varieties, with each variety helping to make a harmonious and complex blend. This can perhaps be compared with cooking, where every ingredient adds to taste and balance. Examples of well known wines made from a blend of varieties include most red Bordeaux, which are usually made from two to five different varieties, and red Châteauneuf-du-Pape where up to 13 can be used.

There are thousands of different grape varieties; the names of some, e.g. Chardonnay, are very well known. Others are largely unknown. Some varieties are truly international, being planted in many parts of the world. Others are found in just one country, or even in just one region within a country. Many varieties have different names in different countries, and even pseudonyms in different regions of the same country.

1.3 What is a grape variety?

The grape vine belongs to a family of climbing plants called Ampelidacae. The family comprises a number of genera, including the genus *Vitis*, which is the grape bearing vine. This genus comprises a number of species, including that of *Vitis vinifera.* It is worth noting that the members of any species have the ability to exchange genes and to interbreed. *Vitis vinifera* is the European and Central Asian species of grape vine, and it is from this species that almost all of the world's wine is made.

Vitis vinifera has, as we now believe, some four thousand different varieties, e.g. *Vitis vinifera* Chardonnay, *Vitis vinifera* Cabernet Sauvignon. Each variety looks different, and tastes different. Some varieties ripen early, others late; some are suitable for growing in warm climates, others prefer cooler conditions; some like certain types of soil, others do not; some yield well, others are extremely shy bearing. Some can produce first class wine, others distinctly mediocre. Whilst these are all factors of relevance to the grower, the actual choice of variety or varieties planted in any vineyard may well, in Europe, be determined by wine laws. These may stipulate which variety or varieties may be planted in a region, and which varieties must be used to make certain wines. For example red Beaune AC must be made from the variety Pinot Noir. Valpolicella DOC must be made from a blend of Corvina, Rondinella and Molinara. It is worth remembering that most of the varieties that we know have been cultivated and refined by generations of growers, although some such as Riesling are probably the descendants of wild vines. Discussion of the characteristics of individual grape varieties is a detailed topic and is beyond the scope of this book. For further information the reader is referred to the Bibliography.

Box 1.1 Definition of crossing

Crossing: Variety produced by fertilising one *Vitis vinifera* variety with the pollen of another *Vitis vinifera* variety.

Box 1.2 Definition of hybrid

Hybrid: A crossing of two vine species. Illegal in the European Union for quality wine production.

It is possible to cross two varieties and thus produce a **crossing**, itself a new variety. For example the variety Müller-Thurgau (also known as Rivaner) is a crossing bred in 1882 by Professor Müller from Thurgau in Switzerland. It was once thought to be a crossing of Riesling (mother) and Silvaner (father), although the current belief is that the father was the variety Gutedel (also known as Chasselas).

It is important not to confuse the term **crossing** with **hybrid** – see Boxes 1.1 and 1.2.

From any variety breeders can select individual **clones**. Clonal selection is basically breeding asexually from a single parent, aiming to obtain certain characteristics such as yield, flavour, good plant shape, early ripening, disease resistance, etc.

In spite of the tremendous development with clonal selection during the past 20 years, most growers believe that old vines give the highest quality juice. Research on genetic modification of vines is taking place, but at present no wine is produced from genetically modified plants.

1.4 Reasons for grafting

Although nearly all the world's wine is produced by various varieties of the species *Vitis vinifera*, the roots on which the *Vitis vinifera* vines are growing are usually those of another species. Why is this?

Vitis vinifera has already been described as the European species of vine, because of its origin. Other species of grapevine exist whose origins are elsewhere, particularly in the Americas. However, although these other species produce grapes, the wine made from them tastes absolutely dreadful. So *Vitis vinifera* is **the** species from which wine is made.

At the time that the countries in the New World were colonised, it became obvious that wine could be produced in parts of many of them, but *Vitis vinifera* needed to be brought from Europe in order to produce sound wines. Of course, *Vitis vinifera* had no resistance to the many pests and diseases that existed in the New World. Disastrously, during the late nineteenth century, these pests and diseases found their way from the USA to Europe, upon botanical specimens and plant material. The vineyards of Europe suffered initially from mildews, which will be detailed later, and then from a most lethal pest, *Phylloxera vastatrix*, that eats into the roots of *Vitis vinifera*, resulting in the vines dying.

1.5 *Phylloxera vastatrix*

Of all the disasters that have struck the vineyards of Europe over the centuries, the coming of *Phylloxera vastatrix* (reclassified as *Daktulosphaira vitifoliae*, see Chapter 5), first discovered in the Rhône valley in France in 1868, was by far the most devastating. By 1877 it had arrived in Geelong in Victoria, Australia and was found in New Zealand in 1885.

Phylloxera had lived in North America, east of the Rocky Mountains, for thousands of years. Naturally the so-called American species of vines had become resistant, the roots having developed the ability to heal over after attack. As the pest wound its devastating way throughout the vineyards of France, into Spain and elsewhere during the latter years of the nineteenth century, many attempts were made to find a cure. Poisoning the soil and flooding the vineyards were two of the ideas tried, which now seem crazy. Many growers simply gave up, and in many regions the amount of land under vines shrank considerably.

Phylloxera is an aphid that has many different manifestations during its life cycle, and can breed both asexually and sexually. An illustration of *Phylloxera* in its root-living form is shown in Fig. 1.3.

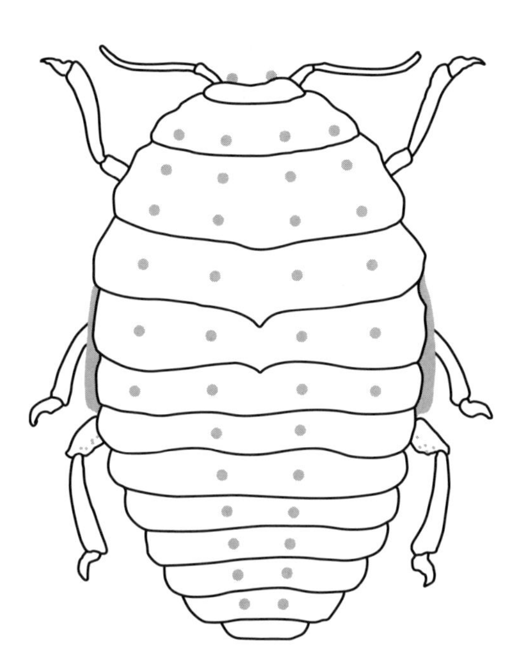

Fig. 1.3 *Phylloxera* louse

When reproducing asexually, eggs are laid on the vine roots and hatch into **crawlers**. These develop into either wingless or winged adults. *Phylloxera* only feeds on the roots of *Vitis vinifera*. The vine's young, feeder roots develop **nodosities** – hook shaped galls. The older, thicker, storage roots develop wart-like **tuberosities**. The cambium and phloem are destroyed, preventing sap from circulating and synthesised food being transmitted downward from the leaves. Root tips can be killed by a process of strangulation by the canker. It must be stressed that *Phylloxera* affects the very life of the vine rather than the quality of the grapes and resulting wine.

As we have seen, American species of vines make unpleasant tasting wine. However, if a *Vitis vinifera* plant is grafted onto an American species rootstock, then the root is resistant to attack, and the plant yields quality grapes perfectly suitable for winemaking. The function of the roots is merely to draw up water and nutrients, including trace elements that will help give the wine flavour. You may know that it is quite possible to produce pears and apples on the same grafted tree, because the design for the fruit lies in the bud.

So the cure to the plague of *Phylloxera vastatrix* was found to be to graft the required variety of *Vitis vinifera* vine stock onto an appropriate American species rootstock. The economic cost of *Phylloxera* is huge – the cost of purchasing grafted vines from a nursery may be four times that of ungrafted material. Although the vast majority of the world's vines are now grafted, there are still several areas where ungrafted vines are common, including Chile, areas of Argentina, and most of the Mosel–Saar–Ruwer in Germany. *Phylloxera* will not live in certain soil types, including sand and slate. The states of Western Australia, South Australia and Tasmania are *Phylloxera*-free. For *Phylloxera* to spread, it is usually transmitted on plant materials, grapes, clothing, tools, picking bins or vehicles – unless transported the aphid will only spread a hundred metres or so in a season. Thus, as in Chile and the *Phylloxera*-free parts of Australia (known as *Phylloxera* Exclusion Zones), a rigorous quarantine programme has excluded the louse, for its arrival in the vineyards could spell economic disaster. To prevent any risk of the aphid spreading on equipment, this must be thoroughly disinfected or heat treated. For example, tractors may be put into a chamber and heated to 45 °C for 2 hours; picking bins may be heated to 70 °C for 5 minutes. New vine planting material must also be heat treated: 50 °C for 5 minutes is effective.

1.6 Rootstocks

There are many different American species of vines, and some of the main ones used as rootstocks are shown in Table 1.1. The choice of species for rootstock will depend not only on the vine stock, but also upon the climate and soil type. Most rootstocks in common usage are hybrids of two American species. For example, rootstock SO4 (suitable for clay–calcareous soils) is a hybrid of *Vitis berlandieri* and *Vitis riparia*, whilst rootstock R110 (a plunging stock

Table 1.1 Examples of species used for rootstocks

Vitis riparia
Vitis rupestris
Vitis labrusca
Vitis berlandieri
Vitis champini

resistant to short periods of drought) is a hybrid of *Vitis berlandieri* and *Vitis rupestris*.

In some regions, grafting of vines still takes place in the vineyards, but bench grafting is the method used in vine nurseries. The grower wishing to establish a new vineyard, or replace existing vines, will order from the nursery a specified clone of a variety of *Vitis vinifera* pre-grafted onto a suitable species or hybrid of root stock. The most common type of graft used nowadays is the machine-made omega cut (see Fig. 1.4).

Some rootstocks that were once considered to be *Phylloxera* resistant (e.g. rootstock AXR) have been succumbing to attacks of the louse during the past 15 years – expensive lessons have been learned by many a Californian grower. It is also argued that during the past decade or two there has evolved a new, even more trouble-some biotype of *Phylloxera vastatrix*.

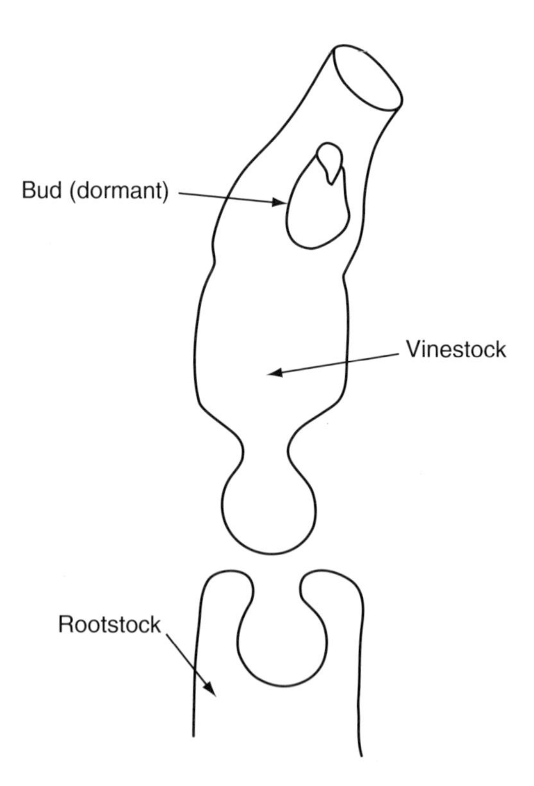

Fig. 1.4 Graft – omega cut

1.7 The lifespan of the vine

Grapevines can live for over 100 years and it is generally accepted that older vines give particularly good fruit. However, with greater age the yield decreases. Accordingly, some producers decide to replant vineyards when the vines have reached a certain age – perhaps 30 to 40 years or so. In areas where there are problems with virus, a vineyard could be grubbed up when the vines are just 20 years old. In the best vineyards, producers will often replace diseased or dying vines on an individual basis, to retain a high average age. Ungrafted vines can have the longest life of all.

Climate

Climate and weather have a major influence on the ultimate quality and style of wine produced. Most of the world's major wine-producing regions are to be found between 30° and 50° latitude in both northern and southern hemispheres. Within this 'temperate zone', grapes will usually ripen satisfactorily. However, there is a wide range of climatic variation. For example, compare the cool, damp and sometimes humid conditions of Northern Europe with the conditions of the extreme south of Europe and North Africa, where grapes struggle to ripen in the dryness and heat. The overall climate of any vineyard region can be referred to as the **macroclimate**. However, as we shall see, climate does vary within areas of any region, and even small variations have an important impact.

2.1 Climatic requirements of the grape vine

2.1.1 Sunshine

Ideally, the vine needs a minimum of 1400 hours of sunshine per year, with a minimum average of 6 to 7 hours per day during the growing season (April to October in the northern hemisphere and October to April in the southern hemisphere). Regions with too much sunshine and heat tend to yield wines that are coarse with high levels of alcohol. Too little sunshine leads to unripe grapes, which in turn result in wines that are over high in acidity but light in alcohol. The vine produces its food by the biological process of photosynthesis, using light. However, for grape growing, the vine needs sunshine more for its heat than light.

2.1.2 Warmth

Growth only takes place at temperatures above 10 °C (50 °F). For the vine to flower successfully in early summer, a temperature of 15 °C (59 °F) is needed. The period from flowering to harvesting averages 100 days, but could be as short as 80 or as long as 150 days.

2.1.3 Cold winter

The vine needs a rest so that its growth is inhibited. Without a winter rest, the vine might yield twice a year and its life would be shortened. Winter frost can be beneficial for hardening wood, and perhaps kills fungal diseases and insect pests. There are, however, vineyards in Brazil where five crops are obtained in a two year period – the climate is hot all year round, and the growing seasons are simulated by turning the irrigation on and off. The vines grown in this way quickly become exhausted and need replacement.

2.1.4 Rainfall

Obviously, vines need water to grow. An annual rainfall of between 500 mm and 700 mm is perhaps ideal; otherwise irrigation will be necessary. Ideally, rain should fall mainly in winter to build up water reserves underground. Rain in early spring helps the vine's growth; in summer and early autumn it swells the fruit. In Europe, nature normally provides sufficient rainfall, but in New World countries irrigation is commonplace. For example, the largest wine producing region in South America is Mendoza in Argentina, parts of which have just 10 mm of rain a year. The area is technically a desert, but the vine flourishes by the use of flood, furrow, or drip irrigation.

Just before harvest, a little rain can help increase the yield, but needs to be followed by warm sun and a gentle drying breeze if undesirable rots are to be prevented. Heavy rain will usually split berries, causing severe damage and diluting juice.

2.2 Climatic enemies of the grape vine

2.2.1 *Frost*

Severe winter frosts can damage vines. Temperatures below –16 °C (3 °F) can kill the vine by freezing the sap, resulting in roots splitting. Frost at budding time can destroy the young buds and shoots; in cases of great severity, an entire crop can be wiped out as a result. Known areas for frost (frost pockets) are best avoided for planting.

However, protection can be given in various ways. In winter, the base of the vine's trunk can be earthed up to prevent damage to the graft. Various measures can be used to mitigate the effects of spring frosts. Traditionally, oil burners were widely used in the vineyard at budding time to circulate air, but these are now considered primitive. Wind machines that circulate air can prove effective. Many producers choose to use an 'aspersion system' for frost protection. This involves installing a sprinkler system which sprays water onto the buds before the temperature falls below 0 °C. The theory is simple: water contains heat, and when the water freezes much of this heat will go into the bud, which will then be protected in its own igloo of an ice pellet. No frost prevention system is always or completely effective.

2.2.2 *Hail*

Hail can do immense damage over relatively small, localised areas, resulting in direct physical damage to both vines and fruit. This can range from scarring of leaves, bruising or breaking of young shoots (the effects of which can carry on to the following season), to the splitting of berries. Split or smashed berries are susceptible to rot or may start to ferment on the vine, resulting in whole bunches being unusable. Various forms of protection include fine netting, either as an overhead canopy or vertical nets against the trellis system. More controversially, the firing of rockets carrying silver nitrate into clouds may cause the ice to fall as rain. For growers in affected areas, insurance is an expensive option.

2.2.3 *Strong winds*

These can have a dramatic effect by damaging canes, breaking shoots and removing leaves. Spring sun and severe wind can be a devastating combination for young shoots and leaves or very young vines. If prevalent when the vine is flowering, then poor pollination and a reduced crop can result.

The detrimental effect of strong winds may be a particular problem in valleys, which can act as funnels. For example, on the steeply terraced hillsides in the northern Rhône valley, vines have to be individually staked, whilst in the flatter southern Rhône, rows of conifer trees have been planted to break the destructive force of the Mistral wind. In other areas, forests and mountain ranges offer some measure of protection.

2.2.4 *Excessive heat*

Heat stress can be harmful to the vine. When there is excessive sun and temperatures exceed 40 °C (104 °F), the vine can shut down and photosynthesis ceases. Thus no more sugars are formed and ripening stops. Grapes can be sunburnt and scarred in the hot afternoon sun.

2.3 Mesoclimate and microclimate

There is often confusion between these two terms. Mesoclimate refers to the local climate within a particular vineyard or part of a vineyard. Microclimate refers to the climate within the canopy of leaves that surrounds the vine. There are several factors that influence the mesoclimate of any vineyard area.

2.3.1 *Water*

Nearness to water, whether rivers, seas or lakes, can bring the vines the benefit of reflected heat. Water can act as a heat reservoir, releasing the heat stored by day during the night. This has the double advantage of moderating temperature and reducing risk of frost. Water also encourages mists and high humidity which can lead to mildews (always unwelcome) or rot (occasionally sought if the production of sweet white wines is the object).

2.3.2 Altitude

For every 100 metres of altitude the mean air temperature decreases by approximately 0.6 °C (1.1 °F). Thus a grower may choose to plant varieties that prefer cooler climates at higher altitudes, e.g. Sauvignon Blanc.

2.3.3 Aspect

Where a vineyard is not completely flat, the direction, angle and height of the slope are important and could, for example, provide protection from prevailing winds. Frost at budding time is less likely to be a problem since it tends to roll down slopes. It is perhaps worth reflecting on some basic botany: the vine sucks in **carbon dioxide** (CO_2) from the atmosphere and combines it with **water** (H_2O) from the soil to create **carbohydrate** ($C_6H_{12}O_6$) in the form of grape sugars. This process is known as assimilation – there is spare **oxygen** (O_2) left over which is transpired to the atmosphere. In the northern hemisphere, the further a vineyard is situated from the equator, the lower the sun is in the sky and the greater the advantage of a south-east facing slope. The vines are inclined into the morning sun when carbon dioxide in the atmosphere is at its highest level, resulting in the manufacture of more sugars.

2.3.4 Woods and trees

Groups of trees can also protect vines from strong winds, but can have the unwelcome effect of encouraging high humidity. They also decrease the diurnal temperature range and can reduce air temperature.

2.4 The concept of degree days

A system of climatic classification based on degree days was devised by the Viticultural School of the University of California at Davis as a result of their research in matching vines to climate. A calculation is based on the seven months' annual growing season according to the hemisphere:

- *Northern hemisphere*: theoretical growing season = 1st April to 31st October
- *Southern hemisphere*: theoretical growing season = 1st November to 30th April.

The vine does not grow at temperatures below 10 °C (50 °F). Each day in the growing season that the mean temperature is above 10 °C is counted and totalled. Each one degree of mean temperature above 10 °C represents one degree day, e.g.:

- *1st May*: mean temperature 14 °C, therefore 1st May = 4 degree days
- *31st July*: mean temperature 24 °C, therefore 31st July = 14 degree days.

Degree days are measured in either Fahrenheit or Celsius, but it is important when comparing different regions to compare like with like, i.e. °C with °C, °F with °F. The total number of degree days is then calculated. For example, in the USA the coolest climatic region (Region 1) totals less than 2500 degree days (F), whilst the hottest (Region 5) has more than 4000. Based on this classification, suitable varieties are selected to match a region's climate. Some viticulturalists challenge the concept of degree days as an appropriate measure of mesoclimates.

2.5 Impact of climate

Let us compare a Riesling wine from a cool climate such as Mosel–Saar–Ruwer, with an alcohol content somewhere between 7.5% and 10% ABV, and a Riesling from Clare Valley, South Australia,

Table 2.1 Comparison of annual sunshine and temperatures

	Duration of sunshine (hours)	Average annual temperature	
		°C	°F
Germany (average in wine regions)	1400	10	50
Bordeaux	2000	12.5	54.5
Châteauneuf-du-Pape	2800	14	57.2

with at least 12% alcohol. Generally speaking, as sugar levels increase during the ripening process, acidity levels fall. Consequently the German Riesling is likely to have higher acidity than the Australian wine, although it should be noted that the Australian producer is permitted to add acid during the winemaking process if it is thought desirable. Unless a derogation is given by the European Union, as happened in the exceptionally hot year of 2003, acidification is not permitted in the cooler regions of Europe. Black grapes need more sunshine and heat than white grapes, to ensure the physiological ripeness of the tannins in their skins. It follows, therefore, that a higher proportion of white grapes are grown in cooler regions; for example, Germany, Alsace, Loire Valley and England. A comparison of the sunshine hours and temperatures of the production regions of three very different styles of wine is shown in Table 2.1.

Climate can be both friend and enemy to the vine. Vines perform at their best where there is a dormant period of about 5 months and a growing and fruit-ripening period of about 7 months. Essentially, sufficient warmth and moisture are required to be able to grow, produce and ripen grapes. It is, however, important that these conditions come at the right time in the vine's annual growth cycle.

2.6 Weather

Whereas climate is determined by geographical location and measured in long-term averages, weather is the result of day-to-day variation of those averages. The weather is the great variable in the vine's growing season and its production. Nearly every other major influence is more or less constant and known in advance. In the end, however, it is the annual weather conditions that can make or break a vintage. Weather does not only influence the quality and style of wine produced in any given year, but can also account for considerable variations in the quantity of production.

Soil

Soil is a critical factor in the production of grapes of the right quality for winemaking, and a variety of factors influence the nature and quality of the soils that are used to grow grapes. This chapter looks at the key factors for consideration in terms of the soil suitability for viticulture and grape production.

3.1 Soil requirements of the grape vine

Many types of soil are suitable for successful vine growing with the correct preparation. Essentially, the function of the soil is to provide anchorage, water and nutrients. Vines can and do thrive in the most inhospitable soils and unlikely sites. In considering the soil structure, both topsoil and subsoil have important roles: topsoil supports most of the root system including most of the feeding network. In some regions, such as Champagne, the topsoil is very thin and nutrient additions are required. The subsoil influences drainage, and ideally enables the roots to penetrate deeply in search of nutrients and water.

There are several factors that need to be considered when planning and preparing a vineyard site.

3.1.1 Good drainage

Like any other plant, the vine needs water and will probe deep to try and find it, so the best soil is one with good natural drainage and free from obstacles likely to impede the roots. There is a viewpoint that in poorly drained soils the grapes may be inferior because the plant is taking up surface water, from recent rainfall or irrigation, rather than the mineral rich deep water. When preparing a vineyard a drainage system should be installed if the natural drainage is inadequate.

3.1.2 Fertility

Although it is generally recognised that the vine thrives on poor soil, vines do require an adequate level of nourishment, and the addition of humus and organic matter may need to be done annually. In highly fertile soils, careful vineyard management techniques are needed to control vine vigour.

3.1.3 Nutrient and mineral requirements

Nitrogen is a most important nutrient requirement, essential for the vine's production of green matter. However, the balance is important so that excessive growth is not stimulated. Desirable minerals include phosphorous and potassium. In addition, trace elements such as magnesium, iron and zinc are beneficial. The finest quality fruit will only be produced when the soil is well balanced and contains appropriate nutrients.

3.2 Influence of soils upon wine style and quality

There is still considerable dissension over the influence that soil has on the quality and style of wine. The traditional Old World view-point, still held in some areas, was that vines only yield good quality grapes if the soil is poor and the vine is made to work hard for its nutrients. Some New World producers dispute this, believing that, providing vigour is controlled, richer soils can give good quality. Before planting a new vineyard, a full soil analysis should take place to examine the chemical and biological composition. The pH of the soil has an impact on the style and quality of wine produced. All other factors being equal, vines grown on a high acid (low pH) soil will produce grapes with a lower acidity than those grown on a low acid (high pH) soil.

The most important physical characteristics of different soil types are those that govern water supply.

3.3 Soil types suitable for viticulture

We will briefly consider some of the types of soil found in various wine regions:

- *Chalk*: a type of limestone that is very porous and has the dual qualities of good drainage whilst providing good water retention. The most famous chalk vineyards are in the Champagne region of France. The albariza soil of the Jerez region in southern Spain is valued for its high chalk content. In this hot, arid region, the soil's capacity for absorbing and storing water over the winter period is invaluable, for very little rain falls during the summer months.
- *Clay*: noted for poor drainage, but consequently good on water retention. In some areas, clay is valued for its role as vineyard subsoil, for example in the Bordeaux district of Pomerol. However, clay soils are slow to warm up after winter, delaying the start to the growing season.
- *Gravel*: prevents water from collecting, ensuring that the vine has good drainage. The left bank of Bordeaux (Medoc and Graves) is famed for its deep gravel banks. During the day the gravel will absorb the sun's heat, discharging it back onto the vines at night.
- *Granite*: a hard crystalline rock rich in minerals that warms up quickly and retains heat. Granitic soils provide good drainage and low fertility. The individual styles of the Beaujolais Crus are influenced by granitic soils.
- *Limestone*: a sedimentary rock consisting mainly of calcium carbonate. In limestone soils, the vine has to burrow its roots deep into the fissures to seek water and nutrients. This type of soil is valued in cool viticultural regions, as for example in producing the great wines of Burgundy, where it contributes to finesse.
- *Sand*: grains of quartz, broken down from siliceous rock. This loose soil type, in which it is particularly difficult to store water and nutrients, is not easily cultivated. A positive quality of the

soil is that the *Phylloxera* louse does not infest sand. Colares on the coast to the north of Lisbon is well known for its sandy soils.

- *Schist*: a coarse-grained crystalline rock that is easily split into thin flakes. It is heat-retaining, ideal for the production of rich robust wines, such as the Port wines of Portugal's Douro Valley.
- *Slate*: a hard, dark and slab-like rock made of clay, shale and other elements. It has the advantage of holding moisture, warming up quickly and retaining its heat, then releasing it at night onto the vines. The blue Devon slate found in the Mosel–Saar–Ruwer contributes to the racy character of the region's wines.

3.4 Soil compatibility

Having considered the qualities of just some of the many soils in which the vine can be planted, it is clear that the vine, and in particular the rootstock, should be compatible with the soil. For example, chalk soils, although providing excellent drainage, can be low in fertility, resulting in low vigour vines. When planting vineyards where there is a high level of active lime in the soil (e.g. most of Burgundy and Champagne), it is important to select rootstocks resistant to lime, otherwise **chlorosis** – the inability of the vine to absorb nutrients – may present a problem.

3.5 Terroir

The influence of soil on style and quality is encapsulated by the French term 'terroir'. Although the concept of 'terroir' is controversial, it is essentially a combination of soil with all its various qualities, local topography (altitude, slope, aspect) and the local climate. As a result, the vine can experience different growing conditions within just a few yards in the same vineyard.

The Vineyard

Viewpoints on viticulture will sometimes conflict, for what is sound theory and practice is not always universally accepted, and this dissension helps contribute to the contrasting styles of wine produced in different regions and countries. One of the Old World concepts challenged by many New World producers is that in order to achieve good quality it is necessary to restrict severely the quantity of grapes produced per hectare of land – see Box 4.1.

Box 4.1 Definition of hectare

Hectare: A measurement of land comprising $10000\,m^2 = 2.47$ acres = 2 football fields.

4.1 Vineyard location

The reasons for the historical situation of the world's great vineyards are complex. Producers in the regions will wax lyrically about the outstanding terroirs of the greatest sites – aspect, soil, mesoclimate, etc. Many of these great vineyards are close to rivers, and we have already seen the influence of these upon the vineyard climate. However, it should be noted that the other great advantage of rivers was that they were the means of transport to the markets. With a strong market outside the region or country of origin, good wines would sell well at higher prices. Higher prices mean more investment in the land, and export markets promote higher quality and the interchange of ideas between supplier and customer, so necessary to improve standards.

Today, when deciding on the location for a new vineyard, communication to the markets may be less of a consideration, but reasonable road access is still required. Indeed, in some countries, harvested grapes may be trucked hundreds of miles to the winery. When investigating potential sites for new vineyards, the factors to be considered are many and complex. The researcher will need to consider aspect, orientation, mesoclimate, soil composition and suitability for the proposed grape varieties, availability of water for irrigation, availability of labour, wine laws and any 'appellation system', and security.

4.2 Density of planting of vines

The number of vines planted per hectare will depend on a number of factors – including historical, legal, and practical. In France, in much of Burgundy and parts of Bordeaux, including much of the Médoc, vines are planted at a density of 10 000 per hectare. The rows are 1 metre apart, and the vines are spaced at 1 metre in the rows. Mechanical working takes place by using straddle tractors, which are designed to ride above vines with their wheels either side of a row. A grower will state that by having such dense planting, the vines are slightly stressed – having to fight for their water, and sending their roots down deep to pick up all the minerals and trace elements. Much of the water goes into the structure of the plant, and only a little is left to go into the grapes. The growers believe that this results in concentrated juice. Thus, dense planting helps restrict the yield per hectare, although not necessarily per plant, which is perhaps just as important. One of the problems that can arise with dense planting is that of soil compaction – the wheels of the straddle tractor follow the same path with each pass. In some other areas of Bordeaux, such as the Entre-Deux-Mers, the district between the Rivers Garonne and Dordogne, the plantings are less dense. Here the soils are heavier, and historically two animals rather than one were needed to pull a plough; thus the wider spacing. The replanting of vineyards sometimes takes place on a vine by vine basis, and the average age of the vines might be 40 or 50 years. Thus historical influences live on simply out of practicality, but also there are many growers who are loathe to change traditional practices.

In other parts of France, plantings may be much less dense, e.g. 3000 vines per hectare in Châteauneuf-du-Pape. However, in many New World countries, plantings might be as low as 2000 vines per hectare. Such a density allows for the passage of large agricultural machinery, helps keep humidity down and lets plenty of sunlight into the vines. Open spacing also helps keep plenty of air circulating, and this can help reduce mildews and fungal diseases of the vine. However, there are vineyards in the New World where the planting is just as dense as in France or Germany, for example parts of Aconcagua and Colchagua in Chile.

4.3 Training systems

We should distinguish between vine training and pruning. Essentially, training combines the design and structure of the support system, if any, with the shape of the plant. Pruning is the management of this shape by cutting as required, and is a means of controlling yield and maintaining fruiting.

The training system used may depend on a number of factors, including historical, legal, climatic and practical. There are several ways of classifying training systems, including by the height of the vines (e.g. low); by the type of pruning (e.g. cane); and by the type of any support system used (e.g. trellis).

4.3.1 Main types of vine training

We will briefly consider the more important of the training systems.

Bush training

Bush-trained vines are either free-standing, or supported by one or more stakes. Historically, this was the most practised training system, but bush-trained vines are not suitable for most mechanised viticulture, e.g. trimming or mechanical harvesting. With a compact, sturdy plant, the system is particularly appropriate for hot, dry climates. The vine arms are spur pruned. If the bush is kept close to the ground, the vine will benefit from reflected heat, giving extra ripeness. If the leaves are allowed to shade the grapes, the system can produce grapes with pronounced green or grassy flavours. Vineyard areas where bush training is practised include Beaujolais and parts of the Rhône valley.

Replacement cane training

Vines are trained using a wire trellis system for support. One or more canes run along the wires according to the particular system used. A popular method of cane training is the **guyot** system, in which one cane (single or simple guyot) or two canes (double guyot) are tied to the bottom wire of the trellis. Amongst the classic vineyard areas that practise the guyot system are much of Bordeaux, including the Médoc, and the Côte d'Or in Burgundy. A diagram of a double guyot vine is shown as Fig. 4.1 and a photograph of recently pruned guyot vines as Fig. 4.2.

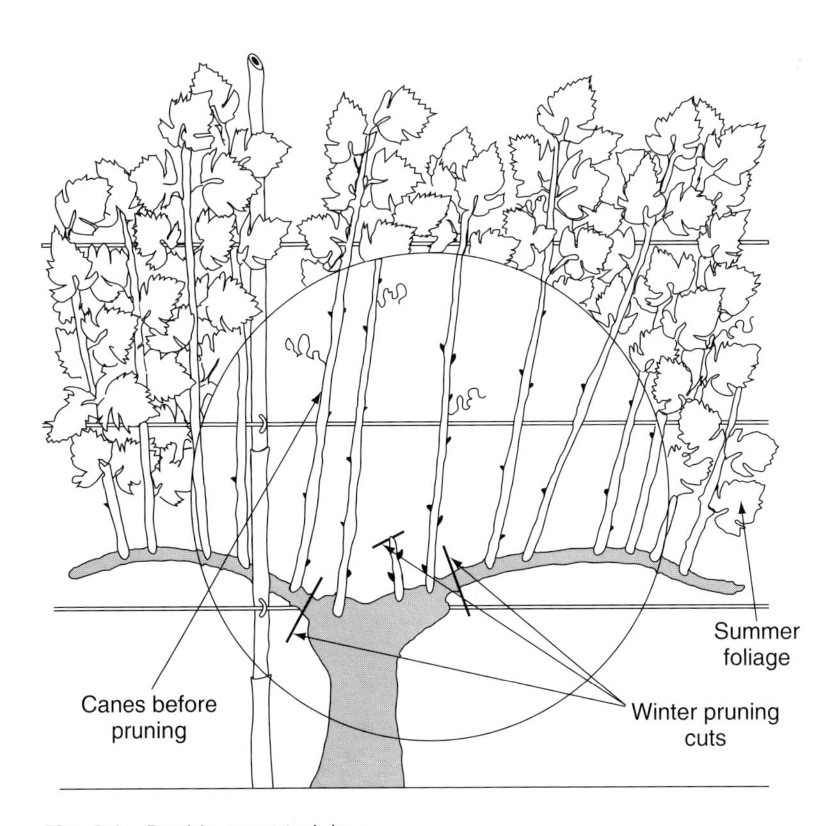

Fig. 4.1 Double guyot training

Fig. 4.2 Vines pruned single guyot

Cordon training

There are many variations of this system, but all have the principle that the trunk of the vine is extended horizontally along wires. The trunk extension (cordon) may be close to or high from the ground. Cordon systems are well suited to mechanisation, including mechanical pruning and harvesting. They are also appropriate in vineyards that suffer from strong winds, which would break the canes of vines trained in a replacement cane system such a guyot.

Vertical shoot positioning (VSP)

This training system requires four pairs of wires on a trellis. As the name implies, the vine's shoots are trained vertically. There are many variations of the system, including both replacement cane and cordon and spur pruning. The latter is extensively practised in many

New World countries, including Australia and Chile. Two cordons are trained in opposite directions along the bottom wire (which is approximately 1 metre off the ground) and each of these will be allowed perhaps ten or twelve spurs which are pruned. With the replacement cane system, practised in New Zealand, there are four canes, two in each direction, with one cane tied to the bottom wire and another to the next wire, which is perhaps 15 cm higher.

Advantages of VSP training include reduced risk of fungal diseases, as the vines are not close to the ground and the canopy is not dense, resulting in less humidity within the canopy. The fruit is contained in a compact zone, with the tips of the shoots higher. Thus mechanisation, including mechanical harvesting, is facilitated. Fig. 4.3 shows unpruned VSP vines in winter, and Fig. 4.4 vines after pruning.

4.3.2 Other training systems

There are many other training systems. Some have been used since ancient times – these include the use of pergolas and vines growing

Fig. 4.3 Vines trained VSP – before pruning

Fig. 4.4 Vines trained VSP – pruned

up trees. Others have been recently developed or refined, and constant research is taking place in schools of viticulture.

4.4 Pruning methods and canopy management

Whatever system of training is used, the vine will need pruning to maintain the shape, manage the vigour and control the yield. The main pruning normally takes place in mid to late winter – in the northern hemisphere, January until early March. Most of the previous season's growth will be cut away. In the case of a replacement cane system, just one or two canes with perhaps 6 to 12 buds each will be left. Bush and cordon systems will be pruned to a number of short spurs with perhaps two buds each. The canes or spurs will be wood that developed in the previous growing season. Grapes will only develop from the flowers that grow on the shoots from the previous year's wood.

If vines are pruned hard the yield is reduced and, as we have seen, the conventional view is that this promotes quality. In France, the more illustrious and precise the appellation, the more restricted

Box 4.2 Definition of hectolitre

Hectolitre: 100 litres (equivalent to 133 standard 75 cl bottles).

is the legally permitted yield. For example, the basic permitted yield for Grand Cru Burgundy is 35 hectolitres per hectare (35 hl/ha).

Many growers challenge this traditional view and claim that high yields can give good quality. There are parts of Australia, heavily irrigated, where the juice that goes into some popular, inexpensive brands is produced at a rate of 200 hectolitres (see Box 4.2) per hectare (200 hl/ha) or more! However, these growers stress that good canopy management, control of the vines' foliage, is crucial to let plenty of sunlight into the vines, which in itself promotes quality.

4.5 Irrigation

In New World countries, irrigation is extensively practised. In many regions, viticulture would not be possible otherwise, but it can also be used as a means of yield enhancement. The use of **flood irrigation**, where water is run over the entire surface of the vineyard, is regarded by many as inefficient, but proponents say that it encourages a large root system, and helps to control certain soil pests. **Furrow irrigation** directs the water more specifically along the rows of vines with the benefit of less water usage. **Micro-jet irrigation** involves the vines being watered by tiny rotating nozzles resembling miniature garden sprinklers, placed above the trellis system. However, the system preferred by most is **drip irrigation**: a plastic pipe, often attached to the bottom wire of the trellis, dispenses a tiny amount of water directly to the base of each individual vine. Restrictive irrigation can be used as a means of vigour control – there is a view that vines, like humans, work best when slightly stressed. Irrigation should not take place in the weeks immediately preceding the harvest.

4.6 The growing season and work in the vineyard

Winter pruning is one of the most labour intensive activities in the vineyard, but throughout the season vineyard operations are necessary. Let us consider the growing season in a cool climate, northern hemisphere vineyard.

Following the pruning in January or February, the vines may need tying. If the guyot system is used the canes will be secured to the bottom wire of the trellis system. Many growers leave this operation until March, when the sap is beginning to rise, and the canes are more flexible. Tilling of the earth between the rows may take place at this time. Some growers like a weed-free vineyard; others encourage growth of some cover crops such as barley or legumes between the rows to help maintain soil balance. These should not be allowed to grow too high, perhaps 60 cm at most, as they would deplete the soil and create high humidity and thus encourage mildews.

The vine will normally bud in April. Any frost now could be disastrous, wiping out much of the year's crop. As we have seen when discussing climate, oil burners, wind machines or aspersion techniques may be used. The last is now the preferred technique in many areas.

During May the shoots will grow rapidly. It is in this month that spraying against mildews begins. Some growers spray routinely, others spray when their early warning system (e.g. a weather station) indicates that mildews might be expected. In June, fairly still, warm weather will encourage a successful flowering. The flower is hardly the most exuberant of flowers – the vine saves its beauty until the wine is in the glass. If the weather at flowering time is cold or rainy, the vines may suffer from **climatic coulure**, the failure of flowers to develop into grape berries due to them not opening or an otherwise unsuccessful pollination. Such weather conditions can also lead to **millerandage**, also known as 'shot berries', where some berries in a bunch remain tiny whilst others develop normally.

By July the fruit should have set, and the berries begin to swell. For the first two weeks or so the berries are growing because the cells are multiplying, but then the growth is due to cell enlargement. A little light rain from late June through to early August is beneficial, for over-stressed vines will produce under-ripe fruit. During the summer, the tops of the vines may be trimmed and leaves and entire shoots may be removed to maximise the vine's efficiency. Lateral shoots should always be removed as these are feeding on the energy of the vine. However, too much leaf removal will result in under-ripe grapes, for the leaves are the food factory of the vine. A formula which allows the amount of leaf area required to ripen a given quantity of grapes to be calculated has been developed by Dr Richard Smart, who is regarded as the world's expert in canopy management. The leaf area required is between 7 and 14 cm^2, dependent upon light and climatic conditions, per gram of ripe fresh fruit. During the summer, green harvesting or crop thinning may also take place.

In August comes the important stage of **veraison**. This is the beginning of ripening. White berries become more yellow or translucent due to the formation of **flavones**, and black grapes take on their colour due to the formation of anthocyanins. From this time the grower may remove leaves from around the bunches to allow the grapes to bask in the sun. The afternoon sun is generally hotter than that of the morning, so it is sound practice to remove leaves only from the eastern side of the vine to minimise the risk of sunburn.

In a cool climate the grapes may not be ripe until October. Many good growers delay their harvest in the hope of a little extra ripeness, but if autumn rains set in this decision may be rued. Whether the grapes are picked by hand or machine will depend on a number of factors including (in parts of the European Union) wine laws: grapes that legislation states must be picked by hand include those for Beaujolais, Châteauneuf-du-Pape, Sauternes and Champagne. The method and timing of picking, along with the other viticultural decisions that a grower has to make, will impact upon the style and quality of the wine. There are advantages and disadvantages to each method and each has its own characteristics. These will be discussed in Chapter 7.

Following the harvest the vineyard will need tidying, stakes and wires replaced, etc. In areas where irrigation is permitted, a post-harvest irrigation is commonplace, to stimulate growth and a replacement of plant energy.

Pests and Diseases

As with any other plant, the vine is susceptible to many types of pest and disease. Some pests, such as parasites, will derive their food from the host without causing too much harm. Others can be devastating. Diseases usually attack during the growing season and, unless controlled, can have a drastic effect on grape quality and the health of the vine. Consequently, growers need to be vigilant and face a constant battle. Most vine diseases need heat or humidity, or both, to flourish. The modern trend is to develop an **integrated pest management** programme that takes into account the environment and aims to reduce the use of pesticides, and this will be discussed in Chapter 6.

5.1 Important vineyard pests

The following list of vineyard pests is far from exhaustive, but includes many of the most serious. Some pose a problem in just a few regions within countries; others are present throughout the viticultural world.

- *Phylloxera vastatrix* (also known as *Phylloxera vitifoliae* and now, due to reclassification, more properly known as *Daktulosphaira vitifoliae*): this is the most serious of all pests and totally destructive, as we have already seen. The only reliable solution is to graft a variety of *Vitis vinifera* onto American species rootstock or a hybrid thereof.
- *Margarodes vitis*: this is the most important of several species of *Margarodes*. It is present across large parts of South America, being the major pest of the grape vine in Chile, and also found

in South Africa. It was first recognised around 20 years ago. Like *Phylloxera* (for which it can be confused), *Margarodes* feeds on vine roots, and results in stunted growth, leaf discolouration, leaf death, and low or no crop. Over time, vine death is inevitable. However, if the vines are encouraged to root to a depth of 3–4 metres, a depth at which the pest cannot survive, prolonged, if stunted, survival of the vine is possible. Heavy rainfall or flood irrigation can reduce pest numbers. Some producers are now grafting to try and combat the pest, but American rootstocks may not be resistant as they are with *Phylloxera. Margarodes* can have the positive effect of restraining vine vigour.

- *Grape moths* e.g. pyrale (*Sparganothis pilleriana*), cochylis (*Eupoecilia ambiguella*), and eudemis (*Lobesia botrana*): the caterpillars of these flying insects attack the buds of the vine in the spring and later damage the grapes, allowing disease to set in.
- *Nematodes*: microscopic worms that feed on the root system, causing a variety of problems for the vine, including water stress and nutritional deficiency. Some species can spread virus diseases through the vineyard. In areas where nematodes are a problem, growers should choose nematode-resistant rootstocks.
- *Spider mites*: these belong to the Tetranychidae family and are mites rather than true spiders, having no separate thorax and abdomen. Species include the red mite and common yellow mite. They infest the leaves and restrict growth.
- *Grape erineum mites* (grapeleaf blister mite): these live on the underside of leaves, causing galls that affect the growth of young vines.
- *Rust mites*: these belong to the same species as grape erineum mite, but are generally found on the upper surface of the leaves. They cause a reduction in yield. It is interesting to note that mites were largely unknown in vineyards before the 1950s, when the introduction of certain insecticides destroyed the previous biological balance.
- *Animals*: these include rabbits, hares, deer, wild boars, and kangaroos which dig, burrow or gnaw, and can be very destructive. Rabbits and hares can strip bark and leaves, whilst larger animals are partial to the grapes.
- *Birds*: can be a serious problem during the grapes' ripening, stripping vines of fruit, especially where vineyards are fairly isolated. Birds are selective and begin their destruction as

soon as the grapes start to ripen. Some growers choose to remedy this by netting their vines or using audible bird scarers.

- *Snails*: mainly a problem in early spring, especially where conditions are wet. They will eat buds and foliage but also encourage birds.
- *Vine weevils*: the adult weevils drill holes into the canes and lay eggs. A legless grub then tunnels along the canes, seriously reducing the strength of the vine.

5.2 Diseases

The vine can succumb to many diseases, the most important of which are considered below.

- *Botrytis cinerea*: this fungal infection may be welcomed or otherwise depending on circumstances. As grey rot it is most undesirable and thrives in wet, humid conditions on vigorously growing vines. It covers the leaves and the grapes with a grey mould. Grey rot will result in off flavours in wine. However, its effect on black grapes is greater than white because it causes loss of colour, tannins and flavour. Outbreaks at harvest time (often induced by rain) can wreck a vintage.

  In its benevolent form, *Botrytis cinerea* occurs as '**noble rot**' on some varieties of white grapes when certain climatic conditions occur. The ideal conditions are damp, misty early autumn mornings giving way to very warm, sunny afternoons. Some of the great sweet white wines of the world are produced from grapes affected by 'noble rot'.

  The grape varieties Sémillon, Riesling and Chenin Blanc are particularly susceptible to noble rot. The fungus gets to work by attacking the inside of the ripe grapes. It consumes sugar, but more water, thus concentrating the sugars and effecting chemical changes. The fungus then attacks the skins, which become thin and fragile and take on a brown/plum colour. The grapes are then known as 'fully rotted'. In a further stage, the grapes will dry up and appear wrinkled. This is the state of perfection, and is known as **confit**. The individual berries within a bunch may well be at different stages of attack. Thus the pickers (usually local people who are experienced at selecting the desired berries) have to return to the vineyards several times in a series of successive pickings in order to select the confit grapes. This process is, of course, extremely costly. The grapes must be picked when dry, and the harvest sometimes continues until late November in the northern hemisphere. The quality and quantity of vintages varies considerably – the crop is at the mercy of the weather, and in some years *Botrytis cinerea* does not develop successfully.

- *Powdery mildew, also known as oidium (Uncinula necator)*: this fungal disease reached the vineyards of Europe from North America between 1845 and 1852. It causes pale grey spores to grow on leaves of the vine and it can overwinter inside the buds. Eventually the grapes will split and shrivel. The disease thrives in mild, cloudy conditions and especially where there is a dense, shady canopy of leaves. It may be treated by dusting the vines with sulphur powder.
- *Downy mildew, also known as peronospera (Plasmopara viticola)*: this is another fungal disease that came from North America around 1878, probably on vines being imported as grafting stock to combat *Phylloxera*. The disease thrives in warm, wet weather, common to many North European wine regions. At least 10 mm of rain in a 24 hour period is necessary for there to be an outbreak. Dense white growth forms on the underside of young leaves. When attacking the grapes, it causes the berries to shrivel up and turn leathery. It may be treated by contact or systemic methods: a common contact treatment is spraying the vines with 'Bordeaux mixture' – copper sulphate and lime. Vineyards thus treated often show a blue hue, and the sprays even give a blue tint to pebbles or gravel on the soil surface.
- *Phomopsis (Phomopsis viticola)*: a particular problem in New Zealand, but found in other cool regions, this disease spreads in wet springs and manifests as black spots surrounded by yellow rings on leaves. Black cracks appear at the base of shoots, which can then break off.
- *Black spot, also known as anthracnose (Elsinoe ampelina)*: this is another fungal disease that needs wet weather and can attack young foliage.
- *Eutypa dieback (Eutypa lata)*: this fungal disease usually attacks older vines, entering through pruning wounds, and results in vine arms dying.

- *Pierce's disease*: prevalent in California, this bacterial disease caused by the organism *Xylella fastidiosa* is transmitted by small winged insects called 'sharp-shooters', which hop from leaf to leaf. The bacteria affects food crops other than grapes, e.g. lucerne, and is commonly associated with reeds and other water vegetation.
- *Viral diseases*: the visually attractive red vine leaves seen in many vineyards in autumn can be a sign of virus-ridden vines. This type of disease is spread mainly by taking cuttings from infected plants. Once the need for grafting rootstocks to combat *Phylloxera* had been established at the end of the nineteenth century, it became critical to check that both the rootstock and scion used were virus-free. Grafting doubles the risk of virus infection as two different plants merge. Viruses are responsible for a number of disorders in grape vines, e.g. fan leaf (court-noué), leaf roll, corky bark and yellow mosaic. Viruses do not usually kill the vine, but gradually reduce growth and yield. As there is no cure for viruses, prevention is the only answer. Nurseries supplying growers must ensure that material is virus-free. One way of ensuring this is to use heat treatment on original mother-garden material. In order to keep their vineyards free of viruses, growers must buy certified stock.

5.3 Prevention and treatments

Spraying at optimum times in the growing season is carried out to minimise or prevent disease and control weeds. Tractors are usually used for this, but small planes or helicopters are an alternative. If, however, the terrain is very steep, then spraying has to be undertaken manually. Spraying is a costly business. Spray treatments may broadly be categorised as either contact, where the treatment remains on the surface of the leaves and other areas, or systemic, in which the agent enters the system of the plant and is freely transported through the tissues. The first spraying, perhaps using lime–sulphur (contact), may start in the spring as the buds swell and soften. In early summer, foliage spraying is often done to prevent fungal infection and mildews forming.

Environmental Approaches in the Vineyard

Numerous environmental considerations must be taken into account and managed appropriately in the growing of vines and production of grapes for wine. Of growing significance is the production of grapes using environmentally sensitive methods and particularly those encompassed by the developing concept of environmental sustainability.

6.1 Integrated pest management (IPM)

The days when grape growers were held hostage to the agro-chemical salesman are perhaps forever behind us. It seems that every grower is practising 'la lutte raisonnée', 'integrated pest management', 'sustainable viticulture' or whatever title they choose to use. During the 1960s, '70s and '80s it seemed that you could not enter a vineyard without seeing routine preventative spraying, or applications of yield 'enhancing' chemicals. Now, cover crops abound and the gentle hum of 'beneficial' insects is the main sound to be heard. And yet the standards of vineyard hygiene are perhaps higher than ever.

In 1991, the ITV (Centre Technique Interprofessionnel de la Vigne et du Vin) published *Protection Raisonnée du Vignoble* (ITV, 1991). It has become both bible and practical guide to thousands of growers needing to fight pests and maladies of the vine, be they cryptogams, bacteria, viruses or mycoplasmata. Conversion to integrated pest management (IPM) is reasonably straightforward. The grower has to identify possible problems, consider the various prophylactic methods of treatment, estimate the risk of the problems occurring (this is the real key to success) and then undertake treatments only as necessary and mindful of the wider consequences. For example, when spraying is necessary, instead of cannon spraying it is possible to use a precise jet machine, or one with panels that collect and recycle surplus liquid, thereby reducing the actual dose per hectare and chemical residues in the soil.

In many regions, networks of weather stations have been established, producing daily reports warning subscribing growers if conditions are such that outbreaks of particular diseases are likely. Prevention might be achieved simply by opening up the canopy, but if spraying is necessary it will be timely, and therefore most effective.

The problem of some insect pests, particularly grape moths, can be reduced by hanging small capsules containing pheromones every 2 metres on the trellis wires. These create 'sexual confusion' in the male, and thus prevent breeding. Beneficial insects are unaffected.

6.2 Organic viticulture

Of course, all farming was once 'organic', but the name was given as recently as 1946, by J.I. Rodale. He created a demonstration farm in Pennsylvania and founded Rodale Press.

However, as little as 20 years ago, there were few producers of organic wine grapes of high quality. Today, many believe that organically grown grapes can actually be of higher quality and produced at lower cost than those dependent upon chemical fertilisers, herbicides and pesticides. Organic viticulture in regions with warm, dry summers is less challenging than in cooler, more humid areas. Nevertheless, there are always pests that pose particular problems. At one time countered with chemicals, these can now be dealt with using natural ingenuity. By way of example, we will consider some particular problems and solutions as dealt with by some organic producers in Chile.

The **burrito** (meaning 'little donkey'), is a nasty mite that when given a chance munches its way through the leaf structure, starving the plant of nutrients and stunting growth. The organic defence is a binding of plastic material with a sticky surface which is wrapped around the vine's trunk about half a metre from the ground. When the burrito tries to climb up to its leafy lunch it gets stuck or turns around and retreats to the earth. Some producers employ a gaggle of geese that eat burritos and anything else they can bury their beaks into as they patrol the vineyard and fertilise the vines. The red (spider) mite is dealt with by the introduction of a larger white spider **Neoseiulus**, harmless to vines, that feeds on the red mite. Grass plantings between the rows of vines complete the Integrated Pest Management (IPM) by giving the white spiders something to munch on. Good vineyard hygiene is particularly necessary in organic production. The problem of **escolito**, an insect that lives on the canes, is reduced by ensuring that all pruned material is taken away and crushed. Wasps are countered by hanging used plastic soft drink bottles into which holes have been punched. The sugary residue attracts the wasps, which are then trapped. Powdery mildew is countered with a spray of sulphur every ten days before veraison (sulphur is a permitted contact treatment) and *Botrytis cinerea* (grey rot) is reduced by sprays of liquified grapefruit, as shown in Fig. 6.1.

Organic vineyard scenes can be particularly bucolic, with geese, ducks, strutting turkeys and even alpacas – all for a purpose. The fowl and alpacas eat surface pests and keep down cover crops. Alpacas are blessed with the attribute of soft hooves, thus reducing soil compaction. Flowering plants are specifically included to attract

Fig. 6.1 Vine sprayed with liquified grapefruit

the vine pests' natural predators. From the creation of a biodiversity project it will take 4 to 5 years for the positive effects to become apparent.

6.3 Biodynamic viticulture

There is little in the world of viticulture that arouses such controversy, scepticism, hostility, or passionate advocacy as the topic of biodynamic farming. Many scientists, including some experts in viticulture, consider biodynamics as 'unscientific tosh', indeed 'black magic'. However, producers of some of the world's greatest wines practise biodynamie. In France these include Domaine de la Romanée Conti, Domaine Leroy and Domaine Leflaive (Burgundy); Zind Humbrecht and Marcel Deiss (Alsace); Chapoutier (Rhône) and Domaine Huet and Coulée de Serrant (Loire).

The birth of biodynamic farming stems from a series of eight lectures entitled 'Spiritual Foundations for the Renewal of Agriculture' given to a group of farmers in June 1924 by Dr Rudolf Steiner. He died a year later, and the term biodynamic, which translates to 'life forces' was not coined until after his death. Biodynamism is a holistic approach which holds that plants are energised not just by the soil but by the air, terrestrial cycles and those of the sun, moon, planets and stars. Of course, synthetic fertilisers, herbicides and pesticides are not used. It should be noted that Steiner formed his ideas before synthetic, carbon-based pesticides were invented.

Biodynamic producers enhance the soil by applying special preparations, at the appropriate time in the natural cycle. There are nine of these, numbered 500–508. For example no. 500 is dung manure, prepared during winter in the horn of a cow. The horn is buried for 6 months, and the contents are then sprayed on the soil at a typical rate of 60 g of manure (dissolved in 34 litres of water) per hectare. No. 502 comprises yarrow flower heads fermented in a stag's bladder. This is then added to compost along with other biodynamic preparations, to help trace elements in the soil become more available to the plants. The compost 'teas' are 'dynamised – stirred both clockwise and anticlockwise, before being added in the tiniest of quantities to the earth. The mixtures will be added on the appropriate day – there are 'leaf days', 'root days', 'flower days' and 'fruit days' according to the passage of the moon – and at the 'correct' time of the day.

Both organic and biodynamic producers are allowed to use copper sulphate and sulphur in the vineyard (and sulphur dioxide in the winery), although some producers are having success with substituting vineyard sulphur use with milk or whey, which counters powdery mildew. Both copper and sulphur are considered essential to life.

There are several certification bodies for organic producers, who thus have a choice as to the organisations to which they may subscribe. One organisation operating in and from the UK is the Soil Association (see Websites section), which is involved in the definition of a variety of organic agriculture certification schemes as well as the auditing of producers to verify compliance with the schemes. As a form of organic agriculture the interests of those involved with biodynamic agriculture are represented by Demeter-International e.V. (see Websites section), which is the main certification body for biodynamic farmers. The organisation operates certification schemes for all kinds of biodynamic agriculture. A rival organisation, Biodivin, is, however, making inroads into the certification of biodynamic viticulture and it operates only in this respect.

The Harvest

This is a critical point during the winemaking cycle when the grapes are picked and transferred to the winery. It is sometimes referred to as 'vintage'. The grower/producer must decide upon the ripeness of the grapes and when to pick. In the northern hemisphere, picking can begin in late August through to late October or even November depending on the mesoclimate and the year's weather. In the southern hemisphere, picking can start in late January through to May.

7.1 Grape ripeness and the timing of picking

The quality conscious grower aims for physiological ripeness, which is not simply a matter of sugar and acidity, but also of tannins and flavour compounds. On black grape varieties, a lot can be determined by tasting the grape and examining the pip, which should be completely brown.

In the weeks and days preceding harvest, the grapes are tested regularly for sugar content. This is usually carried out in the vineyard using a refractometer. Some growers choose to rely on their own sense of taste to make such judgment by going out into the vineyard and tasting the grapes. The timing of picking is one of the most crucial decisions a grower will make during the vineyard year. This decision is usually jointly made between the grower and the winemaker, unless they are one and the same person.

In cooler climates, the grower may be tempted to delay picking in order to obtain a little extra ripeness, especially of the polyphenols. However, the possibility of autumn rain, which may cause rot, can put pressure on timing. Unexpected hailstorms can also be a hazard, causing damage and delay. Where there are different mesoclimates within the vineyard, grapes may ripen at different rates. Of course, different grape varieties ripen at different times. In hotter climates ripening can very quickly accelerate, causing acidity levels to fall rapidly and it is necessary to undertake a rapid picking in order to achieve a well balanced wine.

7.2 Harvesting methods

There are two basic methods of picking: hand and mechanical. In some European regions, wine laws only allow hand picking. Grapes start to deteriorate from the moment they are picked. The method of picking used affects the style and quality of the finished wine. Essentially, the difference is between slow and selective hand picking and the speed of machine harvesting. There are advantages and disadvantages to each method, and the choice may depend on several factors.

7.2.1 Hand picking

This method can be used whatever the training system in the vineyard and on all terrains, however steep or rugged. Of course pickers' labour costs can be substantial, and the team will need housing and feeding. The date of harvest cannot always be accurately predicted well in advance, and so having the skilled labour available when required can sometimes pose problems.

Picking by hand is essentially the cutting of part or whole bunches. For some types of wine made by particular vinification techniques, whole bunches are essential. Of course, the grapes are usually picked along with their stalks. It is possible to be very selective: damaged parts of the bunch or even individual rotten berries may be removed. In vineyards producing grapes for high quality sweet wines such as Sauternes, single berries affected by noble rot can be carefully selected. Gentle handling means minimal damage to the grapes.

Small plastic crates (lug boxes) that stack on top of each other without crushing the grapes are now used by many growers. These enable the grapes to be transferred to the winery in good condition. However, it is still common to see crates that hold 50 kilograms of grapes and, when quality is less important, grapes may simply be dumped straight into the lorry or trailer that will transport them to the winery. Quality conscious producers may employ a team at a sorting table either in the vineyard or at the winery reception in order to eliminate unsuitable grapes.

7.2.2 *Machine picking*

Machine picking has been gaining popularity in the past few decades, although there are many producers who consider even the most advanced machines to be crude and refuse to utilise them. The capital cost is substantial – it is not just the cost of the machines but also the preparation of the vineyard with wide rows and the installation of appropriate trellis training systems. Some growers rely on the hiring of a picking machine, but this may not be available at the required time, especially if the weather in the growing season has been abnormal.

Mechanical harvesters can only be used on flat or gently sloping terrain. The berries are picked by being vibrated off the vines and are collected in a reception hopper. The vibrating arms can cause damage to vines and even the trellis system. It is important that a precise fruiting zone is controlled on the trellis hedge.

Machines can pick a vineyard in a fraction of the time that a hand-picking team needs, which is important where picking time is limited, e.g. bad weather is forecast. Also, mechanical harvesters can operate 24 hours a day and have the advantage of being able to undertake picking at night. This is important in hotter regions, ensuring that grapes are picked at cool temperatures and delivered to the winery in good condition. The winemaker does not want to receive hot grapes. A disadvantage of mechanical harvesting is that the process is unselective – the machine is unable to distinguish between healthy and diseased grapes. The action of machines means that only berries can be picked and not whole bunches. Stalks, which can act as drainage channels in the early stages of winemaking, are not available. Machines do pick MOGs – *M*aterials *O*ther than *G*rape. These can include ladybirds, snails, leaves and vineyard debris.

7.3 Style and quality

Ultimately, the method of picking and handling of the grapes, the timing and the speed of their arrival at the winery, will impact on the style and quality of the finished wine. If the grapes are damaged, oxidation will set in and prolonged skin contact with the juice, particularly for white wines, can lead to excessive phenolics and the loss of aromatics.

Vinification – The Basics

The process of turning grapes into wine is called vinification. It is a highly developed process, full of scientific complexity if one looks deeply enough. It is not the intention of this chapter to explore vinification in detail or engage with difficult concepts, but to draw the reader into the subject.

8.1 Basic principles of vinification

The sugars contained in the pulp of grapes are fructose and glucose. During fermentation enzymes from yeast convert the sugars into ethyl alcohol and carbon dioxide in approximately equal proportions and heat is liberated:

$$C_6H_{12}O_6 \rightarrow 2CH_3CH_2OH + 2CO_2 + \text{Heat}$$

Additionally, tiny amounts of other products are formed during the fermentation process, including glycerol, succinic acid, butylene glycol, acetic acid, lactic acid and other alcohols. The winemaker has to control the fermentation process, aiming for a wine that is flavoursome, balanced and in the style required. The business of winemaking is fraught with potential problems, including stuck fermentations (the premature stopping of the fermentation whilst the wine still contains unfermented sugars), acetic spoilage, or oxidation.

The amount of sugar in must, and the reducing sugar (glucose and fructose) in fermenting wine, can be determined by using a density hydrometer. A number of different methods can be used to measure density, which can be directly related to sugar content using appropriate formulae or, more readily, prepared tables. Different countries tend to use one or other of the methods available. The methods most commonly in use are the Baumé, Brix, Balling and Oechsle methods. Throughout the vinification process it is essential to maintain accurate records, including temperature and gravity readings.

There are important differences in the making of red, white and rosé wines. Grape juice is almost colourless. For red wine, the colour must be extracted from the skins. For white wines, however, some winemakers choose to have a limited skin contact between the juice and skins because it can add a degree of complexity. Of course, white wine can be made from black grapes, commonly practised in the Champagne region.

Rosé wine is usually made from black grapes whose juice has been in contact with the skins for a limited amount of time, e.g. 12 to 18 hours. Consequently, a little colour is leached into the juice.

8.2 Winery location and design

Wineries vary from centuries old stone buildings externally of timeless appearance to the modern practical constructions, perhaps built to the design of a specialist winery architect. The winemaker who has inherited the ancestral château with all its low ceilings, narrow doors and dampness may long for the blank canvas available to those with an embryonic project. Conversely, when faced with a sterile hangar of his or her newly built winery, the producer may envy the neighbour's ancient property, especially in an age when wine tourism and cellar-door sales have become a major contribution to the profits of many businesses. During the past twenty years or so many wineries have been built that have neither walls nor roof, as shown in Fig. 8.1.

When planning a new winery, practicalities usually dictate. Keeping the building cool is a major consideration. In a hot climate, the building should be sited on a north–south axis, with the shorter walls into the midday sun. It can also be beneficial not to have windows on the western side, facing the hot afternoon sun. Adequate insulation is required, for the ideal ambient temperature inside is no more than 20 °C (68 °F). Pumping of must or wine over long distances should be minimised. Some winery designers aim for gravity flow, with grapes arriving at or being hoisted to the top level for crushing. The must or wine moves only by gravity to subsequent stages of the winemaking process. Some wineries have been built

Fig. 8.1 Winery without roof or walls

into hillsides to facilitate the gravity flow principle and minimise the ambient temperature.

Wineries use copious amounts of water, particularly for cleaning purposes. Although water is not normally an ingredient in winemaking, as a general rule, for every 1 litre of wine produced 10 litres of water will be used in the winery. Adequate drainage is required throughout, especially in the vat rooms and areas where equipment washing takes place.

Ventilation is vital. Every litre of grape juice, when fermented, produces about 40 litres of carbon dioxide. The gas is asphyxiating and consequently deaths in wineries can occur.

8.3 Winery equipment

Equipping or re-equipping a winery is very capital intensive, and some of the items are used for just a few days a year. Equipment utilised includes:

- crusher/destemmer(s)
- fermentation and storage vats

- press(es)
- pumps
- fixed and moveable pipes and hoses
- filters
- refrigeration equipment
- barrels, if utilised
- bottling line, if wine is to be bottled on the property
- laboratory equipment
- cleaning equipment.

8.3.1 Fermentation vats

Wine is usually fermented in vats, although barrels are sometimes used, particularly for white wines. Traditionally, wine was made in either shallow stone tubs called 'lagars', in which the grapes would be trodden, the gentlest of crushing, or in open wooden vats. Lagars are sometimes still used in Portugal by some producers of Port wines. Cuboid vats made of concrete (béton) or cement (ciment) became very popular with producers in the early to mid-twentieth century and many still regard these as excellent fermentation vessels. They can be built in a single or double tier against the walls of the winery, utilising all available space. Concrete or cement vats should be lined, for the acids in wine can attack the material. Glazed tiles were often used for this purpose, but they are prone to damage, and wine getting behind them would present a serious hygiene problem. A much better lining material is epoxy resin, which is easily cleaned, and the vats can be re-lined when necessary.

Stainless steel (inox) vats are now commonplace, but it is interesting to note that it is only from the late 1970s that they gained universal acceptance. When, in the 1960s, stainless steel was installed at Château Latour in Bordeaux, the locals accused the British (who owned the property at the time) of turning the illustrious château into a dairy! At the start of the 1970s only three Bordeaux properties were using stainless steel, but now its presence in winemaking is becoming commonplace. The great advantages of the material are easiness of cleaning, and the ability to build in cooling systems. There are two grades of steel used for construction: 304 grade is really only suitable for red wine fermentations, whilst the better 316 grade, which contains molybdenum and is harder and more corrosion resistant, is ideal for both reds and whites. Stainless steel vats may

Fig. 8.2 Winery with assorted vats

Fig. 8.3 Winery with stainless steel vats

be open topped or closed, with a sealable hatch lid and fermentation lock. Variable capacity vats are also available and are useful even when only partly filled. These have a floating metal lid, held in place by an inflatable plastic tyre at the perimeter. Fibreglass vats are occasionally used as a less costly alternative to stainless steel.

Although in the late twentieth century many producers replaced their wooden vats with stainless steel, wood is again gaining popularity as a material for vat construction at some small and medium-sized wineries. The difficulties of maintaining and sanitising the vats are ever present, but there is less risk of the wine being affected by reductivity, as will be discussed in Chapter 17. Illustrations of wineries containing various vats are shown in Figs 8.2 and 8.3.

Red Wine Making

This chapter describes the processing of grapes to make red wine. Various approaches can be, and are, taken in the production of red wine and many different processes may be used. The following processes are usually undertaken, but it should be borne in mind that the list is neither prescriptive nor exhaustive.

9.1 Destemming and crushing

On arrival at the winery, the stalks may be removed to prevent any bitterness tainting the juice. Following this, the grapes can be lightly crushed. Both these tasks can be performed by the same machine, a crusher–destemmer, as illustrated in Fig. 9.1. The grapes

Fig. 9.1 Destemmer, crusher and pump

are fed via a hopper into a rotating slotted cylinder. As this rotates, the berries pass through the slots, leaving the stalks behind – these are then expelled from the machine and can be used for fertiliser. The grape berries are passed through a series of rollers that can be adjusted to give the chosen pressure in order to release the juice.

For certain wines, where whole bunches are required for pressing, e.g. Beaujolais, the grapes will not be destemmed or crushed. There is a growing trend amongst some red wine makers to include at least a percentage of whole grapes in their fermentations.

9.2 Must preparation

The resultant mixture of grape juice with seeds, skins and pulp (must) now has to be prepared for fermentation. Various additions and adjustments may be undertaken.

- *Sulphur dioxide* (SO_2): this is the winemaker's universal antioxidant and disinfectant, and is used at many stages in winemaking. To prevent fermentation starting prematurely, it may be added to inhibit the action of wild yeasts and bacteria. These organisms require oxygen for growth, and are naturally found on grape skins. Wild yeasts (which can cause off flavours) die when 4% alcohol is reached. Naturally occurring wine yeasts found in the vineyard and the winery work without the action of oxygen and thus can work even if the must is blanketed with sulphur dioxide. In many parts of the world, winemakers now prefer to use selected cultured yeasts for greater control, reliability and for specific flavours. It should be noted that producers often speak of wild yeast fermentation, when referring to use of the natural yeasts that come into the winery on the grapes.
- *Must enrichment (chaptalisation)*: in cooler climates, grapes often do not contain enough sugars to produce a balanced wine. This may be addressed by chaptalisation – the addition of sucrose to the must or the juice in the early stages of fermentation. It is important that only the minimum necessary amount is added or further imbalance will be created. The practice is not permitted in many, hotter, countries. The European Union

is divided into zones according to crude climatic conditions, and the amount, if any, of chaptalisation allowed varies according to the zone. In some countries concentrated grape must is used instead of sugar.

- *Acidification*: this may be necessary if the pH of the must is too high, that is, if the acidity is too low. The addition of tartaric acid is the usual method employed. The addition of malic acid is not permitted within the European Union, although it is not uncommon in Argentina. Acidification is generally not permitted in cooler regions of the European Union (classified in the EU as regions A and B), but derogations are possible. For example, in 2003 Regulation (EC) No. 1687/2003 permitted acidification on account of the exceptional (hot) weather conditions.
- *De-acidification*: this may be necessary if the pH of the must is too low. It is not permitted in warmer regions of the European Union. There are a number of materials that may be used, including calcium carbonate ($CaCO_3$), perhaps better known as chalk, potassium bicarbonate ($KHCO_3$), and potassium carbonate (K_2CO_3). Another agent that may be utilised is Acidex®, which is a double-salt seeded calcium carbonate designed to reduce both tartaric and malic acids in must or wine. The product was developed in Germany, where the cool climate often produces grapes of high acidity. NB: Calcium carbonate only reduces the tartaric acid.
- *Yeast*: cultured yeasts may be added, or the winemaker may simply utilise the natural yeasts present on the skins.
- *Yeast nutrients*: as living organisms, yeasts need nutrients, and B group vitamins may be added to promote their growth.
- *Diammonium phosphate (DAP)*: this may be added, usually at a rate of 200 mg per litre of must, to help ensure that all the sugars are fermented out and to stop the formation, during fermentation, of hydrogen sulphide (H_2S), which is most undesirable. Its use is common in New World countries, particularly when the musts are nitrogen deficient.
- *Thiamine*: thiamine (vitamin B_1) may be added in the early stages of fermentation to help increase yeast populations and prolong their life. The yeast *Brettanomyces* is regarded by most as a spoilage yeast and, therefore, is generally undesirable. It needs thiamine to grow, and additions of this need to be undertaken with caution.

9.3 Fermentation, temperature control and extraction

As we have seen, the process of fermentation results in the conversion of sugar by the enzymes of yeast into alcohol and carbon dioxide.

9.3.1 *Fermentation*

The fermentation of red wine takes place with grape solids present, in order to extract colour from the skins. Initially the fermentation can be very tumultuous, but as more sugar is converted, the rate slows down. In the majority of cases the fermentation is continued until the wine is dry or off dry, and depending upon the richness of the must, the final alcohol concentration is generally in the range 11% to 14.5% by volume.

9.3.2 *Temperature control*

The fermentation process is a turbulent one and creates heat naturally. During red winemaking, fermentation may begin at about 20 °C, but temperatures may rise to 30° to 32 °C. Yeast ceases to work if the temperature rises above approximately 35 °C. Therefore some form of temperature control may be necessary, especially in warmer regions, to prevent this happening before the sugars are fully fermented. It is only in the past few decades that winemakers have had the equipment and ability to be able to have real control.

One hundred grams of sugar in a litre of juice has, when fermented, the theoretical potential to heat the juice by 13 °C. If grape must of 11.1° Baumé commences fermentation at 15 °C, the potential temperature increase could be to 41 °C. If fermentation of must with a Baumé of 14.5° commences at 20 °C, the potential temperature increase is to 54 °C. Of course, some of this heat is naturally dissipated during the period of fermentation through the walls of the vat and with the rising carbon dioxide, through the juice to the surface.

Good colour extraction requires warm fermentations. However, cooler fermentations aid the growing of yeast colonies and give higher alcoholic degrees. The warmer the temperature, the less time the fermentation takes. Accordingly, managing the temperature can be quite a tricky exercise. A winemaker may decide to start a vat fairly cool, at say 20 °C, and allow it to rise naturally to around 30 °C to aid extraction. In the latter stages, the vat may be cooled to 25 °C or so, to ensure complete fermentation to dryness. In the cool underground cellars of regions like Burgundy, the temperature of small vats or barrels can be self-regulating. Cooling equipment may be required for larger vats. Wine can be pumped through heat exchangers to reduce (or increase) temperature. Stainless steel tanks are now commonly wrapped with water or glycol cooling jackets. Alternatively, they may be cooled by showers of cold water running down the outside. In concrete or wooden vats a metal cooling device (drapeau) can be inserted or built in.

9.3.3 *Extraction*

The traditional process for red wines is for the grape mass to be fermented in open vats. The solids and skins rise to the surface with the CO_2 and create a floating cap. This is a disadvantage because the skins need to be in contact with the juice for there to be good extraction of colour and tannins. Also, acetic bacteria thrive in such a warm, moist environment, risking spoilage of the juice.

Consequently, during the process, the juice is drawn out from near the bottom of the vat and pumped up and sprayed over the cap to submerge it. This process, known as **remontage**, has the additional benefit of aerating the must, which helps to boost the yeast colonies. This technique of **pigeage**, simply punching down the cap, is used for some varieties, particularly Pinot Noir which needs a very gentle extraction process. Although vats can now be fitted with mechanical pigeage equipment, in many wineries submerging the cap is done by hand with wooden paddles, sticks, or even the feet of the cellar staff precariously suspended over the fermentation vats.

9.4 Maceration

Depending on the style required, the wine may be left to soak with the skins after completion of the alcoholic fermentation, until sufficient colour, flavour and tannins are extracted. This maceration could vary from 2 or 3 days up to 28 days. If an 'early drinking' red is required, the juice may be drained off the skins just before completion of the fermentation. Alternatively, if the winemaker believes that a post-fermentation maceration would not be beneficial, with the risk of hard tannins being extracted, the tank may be drained hot, that is, immediately after the alcoholic fermentation, before the wine has cooled.

9.5 Racking

Racking is the process of transferring juice or wine from one vessel to another, leaving any sediment behind. After the fermentation and any maceration the wine, known as *free-run juice*, will be run off to another vat. The skins and other solids are left behind in the fermentation vat. These will be transferred to the press to obtain further juice. Between 10 and 15% of the total juice comes from the pressing process. Free run juice has less tannin and is usually considered superior to pressed juice. Some producers choose to make their wine only from free run juice. Racking will also take place at various other times in the winemaking/maturation process to remove the wine from lees and sediment, and clarify it. Aeration can also take place during the racking process, and an addition of sulphur dioxide may be made if necessary.

9.6 Pressing

The juice released from the press will naturally be higher in tannin and colouring pigments. More than one pressing may be carried out, but with each pressing the press wine becomes coarser. Some presses are controlled to exert varying levels of pressure at different stages, often gentle at the beginning but harder with each subsequent pressing. There are a variety of types of press available, including basket press, horizontal plate press and pneumatic press. The characteristics of these will be considered later.

9.7 Malolactic fermentation

This usually follows the alcoholic fermentation and so is sometimes referred to as secondary fermentation. Yeast is not involved. It can be described as a transformation caused by the action of strains of bacteria of the genera *Lactobacillus, Leuconostoc* and *Pediococcus*. Harsh malic acid (as found in apples) is converted into softer tasting lactic acid (as found in milk). One gram of malic acid produces 0.67 g of lactic acid and 0.33 g of carbon dioxide. Malolactic fermentation can be induced by warming the vats, or inoculating with strains of lactic acid bacteria. Alternatively, it can be prevented by treating the wine with sulphur dioxide and/or keeping the wine cool. It gives the wine a slight 'buttery' and/or toasty nose and sometimes a certain amount of complexity.

9.8 Blending

This is an important operation in wine production. Once fermentation has finished there will be various vats or barrels containing wines from different vineyards, sections of vineyards, districts or even ages of vine. Individual grape varieties, having been picked according to ripeness, will have been fermented separately. Blending of these various vats will then take place in order to achieve the desired final style and quality of wine. The prime reasons for blending are to have a product that is greater than the sum of its parts, to even out inconsistencies and perhaps to maintain a brand style.

This selection process could involve a great number of component parts. Where different grape varieties are blended together, the winemaker considers what proportion of each variety is included in the blend. Some varieties or vineyard blocks may perform better in a particular season. Where high quality is the object, some vats may well be rejected and used in a lower quality wine or sold off in bulk.

9.9 Maturation

Immediately after fermentation, wines may taste rough and fairly unpleasant. A period of maturation is required during which the tannins soften and acidity levels fall. The choice of maturation vessel and the period of time depend upon the style of wine to be produced and quality and cost factors. There are many types of maturation vessels, including stainless steel vats and wooden barrels. Some wines which are intended for early drinking, such as inexpensive or branded wines, need little or no maturation. Stainless steel is an ideal storage material because it is impermeable to gases such as oxygen. The wine therefore is stored until required for bottling. Of course, stainless steel is ideally suited to temperature control and is used where long-term, oxygen-free storage is required, for example when inexpensive wines are held prior to being 'bottled to order'.

Most high quality red wines undergo a period of barrel maturation – usually somewhere between 9 and 22 months. During the time in barrels, the wine will undergo a controlled oxygenation and absorb some oak products, including wood tannins and vanillin. Barrel size has an effect on the maturation of the wine; the smaller the barrel the quicker the maturation. Temperature also plays an important part; the lower the temperature, the slower the maturation. When the barrels have been filled they will be tapped with a mallet to dislodge air bubbles, which rise to the surface of the wine, thus removing oxygen. During the period in barrel the wine will be racked several times to aid clarification – for example red Bordeaux wines are normally racked four times during the first year's maturation, and perhaps once or twice in the second. Many producers regularly top up the barrels to replace wine lost to evaporation, but there is a viewpoint that if the barrels are securely sealed the ullage at the top is a partial vacuum and topping up may not be beneficial. The processes of preparing the wine for bottling will be considered later.

Dry White Wine Making

Although the techniques used for making white wine are similar to red, the sequence of operations is different.

10.1 Crushing and pressing

10.1.1 Crushing

On arrival at the winery, white grapes should be processed with minimum delay to avoid deterioration and the onset of premature fermentation. In hot climates, ideally grapes are chilled before the crushing stage. In most cases, they will be destemmed and lightly crushed before pressing. It is becoming increasingly popular for the winemaker to allow some hours or even days of skin contact with the juice at low temperatures before the grapes go to the press. This may extract more aroma compounds and enhance flavours.

When white wine is being made from black grapes, crushing must be avoided to prevent any colour being leached into the juice, and whole clusters are sent to the press. Many winemakers like to do this with clusters of white grapes too, believing this gives a great purity of juice.

10.1.2 Pressing

Unlike in the red wine process, pressing occurs before fermentation. Gentle pressing, avoiding the crushing of pips, results in better quality juice. The skins of white grapes are not used during the fermentation process.

10.2 Must preparation

The juice is drained from the press into settling tanks. If the must is not cleared of solid matters, then off tastes can result. A simple settling over a period of 12 to 24 hours, a process known in France as **débourbage**, may take place. The must is then racked to another tank for fermentation. Clarification can also be achieved by using a centrifuge, which speeds up the process, or by filtration. Very fine particles can also be removed by treating the must with bentonite. This is a form of clay earth which acts as a flocculent, attracting and binding fine particles which then settle out of suspension. If the must is over clarified, fermentation may proceed slowly and the wine may not ferment to completion. Yeast nutrients attach to solid matter within the wine, making them accessible to yeasts. The removal of solid matter can reduce the availability of nutrients, with detrimental effects on yeast growth.

The must may then be passed through a heat exchanger to lower the temperature. This not only prevents a premature onset of fermentation, but also preserves freshness and flavours. The must is treated with sulphur dioxide to prevent aerobic yeasts and spoilage bacteria working. Selected cultured yeasts may then be added.

10.3 Fermentation

The must is then pumped directly to the fermentation vats or barrels. With careful winery design, this process can be achieved simply by gravity, for pumping is by nature a harsh process. Selected cultured yeasts may be introduced and are often used in white wine making. White wine is usually fermented cooler than red wine, at around 10° to 18 °C and over a longer period to preserve primary fruit flavours. Each vat is under temperature control, usually with its own chilling system. Cool fermentations are desirable for aromatic whites, with less cool fermentations used when full-bodied wines are desired. Some white wines are fermented in barrels to give particular characteristics to the wines and a better integration of oak flavours. Barrels also give a better supply of oxygen, required by the yeast in the early stages of fermentation. The choice of new or used oak from particular origins will have been made. It may be that different lots are fermented, for example, in new, second fill and third fill oak barrels and blended together at a later stage. After fermentation, the wine may be left on the lees which, from time to time, may be stirred. This operation is known as **bâtonage** and gives a yeasty, and perhaps creamy flavour to the wine.

Vat matured wines may also be left lying on the lees. We should distinguish between the gross lees, the large lees from the fermentation pre-racking, and the **fine** lees that will fall after the wine has been racked for the first time. Ageing on lees may take place on either gross or fine lees, depending on the textures sought. Stirring of lees may also take place in vat – a 'propeller' type of device may be inserted through the side of the vat to rouse the settled sediment. It should be noted that some varieties (e.g. Chardonnay) may benefit with lees stirring, whilst others (e.g. Riesling) will suffer if the lees are roused.

10.4 Malolactic fermentation

Malolactic fermentation may follow alcoholic fermentation to soften any aggressive acidity. Some white grape varieties, e.g. Chardonnay, work well with malolactic fermentation, whereas others may not. Other varieties which are valued for their crisp acidity, such as Riesling or Sauvignon Blanc, do not usually undergo malolactic fermentation. Also, in some countries the crispness of wines that have not undergone the malolactic fermentation is a desirable feature. For example, a Sancerre from the Loire Valley in France, a wine made from Sauvignon Blanc, can be much appreciated for its lively acidity. Following the malolactic fermentation, the wine is racked into clean vats or barrels.

10.5 Maturation

Much white wine is stored in stainless steel or concrete vats until ready for bottling. It is important that oxygen is excluded, and the vats should be kept either completely full or blanketed with nitrogen or carbon dioxide. Even if white wine has been fermented in barrel, it may be that it continues its maturation in barrel, too, to obtain more oak flavours. Barrel and other forms of oak maturation will be discussed later.

Preparing Wine for Bottling

To ensure that wine is of the clarity expected and remains stable in the bottle for its life, various treatments may be undertaken; however, many makers of fine wines believe treatments should be kept to a minimum.

We have already seen that racking is one of the operations necessary to achieve clear wine and to reduce the risk of off flavours. It is important for the wine to be protected from oxidation during the racking process, and vats or barrels may be sparged with inert gases to achieve this. A number of operations may be undertaken to ensure the clarity of wine.

11.1 Fining

With the coarse sediment removed by racking or centrifuge, there remains other lighter matter suspended in the wine known as **colloids**. These are capable of passing through any filter. If not removed they will cause the wine to look 'hazy' and then form a deposit. The colloids are electrostatically charged and can be removed by adding another colloid with the opposite charge. Examples of such fining agents are egg whites, gelatine, isinglass (obtained from swim bladders of fish) and bentonite. Quantities need to be carefully controlled otherwise the fining agent itself will form a deposit, or a further, opposite, electric charge may be created. Fining may also be used to remove excess tannin and so improve the taste of the wine. Phenolic compounds are absorbed by the substance PVPP (polyvinylpolypyrrolidone). This may be used at the fining stage to remove colour from white wines and help prevent browning.

11.2 Filtration

Filtration is the process used to remove solid particles, and may take place at various stages in winemaking, for example must or lees filtration. However, one of its main uses is in the preparation for bottling. The processes of fining and filtration are not interchangeable. Filtration requires care and expertise. Today, some winemakers choose to use minimum filtration to avoid what they see as 'stripping the body' away from the wines.

There are three principal categories of filtration, which may be used at different stages in the winemaking process:

11.2.1 Earth filtration

This filtration method is used for initial rough filtration and can remove large quantities of 'gummy' solids, which consist of dead yeast cells and other matter from the grapes. The filtration takes place in two stages. Firstly, a coarse grade earth called **kieselguhr**, which is commonly used as the filter medium, is deposited on a supporting screen within a filter tank. A mixture of water and kieselguhr may be used to develop the filter bed. This is known as precoating. Secondly, more earth is mixed with wine to form a slurry that is used continuously to replenish the filtration surface through which the wine passes. Wine is passed through the filter and the bed gradually increases in depth. Eventually it will clog and the kieselguhr will have to be completely replaced with fresh material. An earth filter is shown in Fig. 11.1.

The **rotary vacuum filter** consists of a large horizontal cylinder or drum with a perforated screen covering the curved surface. A filter cloth is stretched over the curved surface. The cylinder rotates in a trough which contains initially kieselguhr and water. A vacuum is drawn on the cylinder and the kieselguhr–water mix is drawn onto the cloth. The water is drawn into the cylinder, leaving a layer of kieselguhr on the cloth to act as a fine filter medium. Wine is fed to the trough and filtered through the kieselguhr. As the surface fouls, more kieselguhr can be added to the wine to replace the fouled layer, which is scraped off by blades as the cylinder rotates. A sophisticated filter is found in the totally enclosed rotary vacuum filter, which reduces the risk of oxidation damaging the wine, but with this type

Fig. 11.1 Earth filter

the fouled earth has to be removed manually. Two open vacuum filters are shown in Fig. 11.2.

11.2.2 Sheet filtration (sometimes called plate and frame filtration)

A series of specially designed perforated steel plates are held in a frame. Sheets of filter medium (cloth or paper) are suspended between the plates, which are then squeezed together by screw or hydraulic methods. The filter sheets are available with various ranges of porosity. Wine is pumped between pairs of plates to pass through the filter sheets into a cavity in the plates and then to exit the system. Yeast cells and other matter are trapped in the fibres of the filter media. A sheet filter is shown as Fig. 11.3.

11.2.3 Membrane filtration

If undertaken, this is the final filtration process and used just before bottling. The wine needs to have been well clarified by other filtration

Fig. 11.2 Winery with two vacuum filters

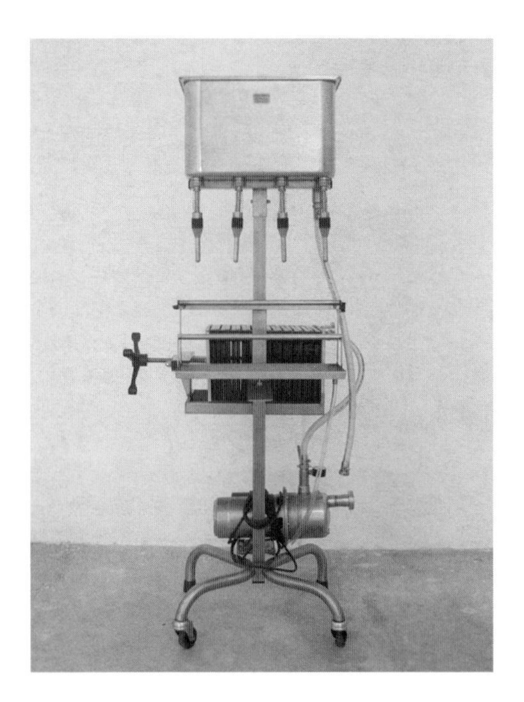

Fig. 11.3 Sheet filter with a four-head bottler

methods before using this process. Membrane filters are constructed of plastic or ceramic materials which are porous to water and compounds of small molecular size, e.g. alcohol, colour and flavour compounds found in wine. Large compounds such as proteinaceous materials will not pass through. Wine is passed under pressure across the membrane surface to filter through, leaving undesirable materials trapped on the membrane. The process is not used for full-bodied red wines as it can reduce body and flavour. A double membrane filter is shown in Fig. 11.4.

Fig. 11.4 Double membrane filter

11.3 Stabilisation

Stabilisation may be carried out to prevent tartrate crystals forming after the wine has been bottled. The tartrates are either potassium or calcium salts of tartaric acid and are totally harmless. They are sometimes found on the cork or as sediment in the bottle, and sometimes cause unwarranted concern to consumers.

To inhibit the precipitation of tartrate crystals in bottle, the wine is chilled to –4 °C, or colder in the case of liqueur (fortified) wines. After approximately 8 days the crystals will have formed, and the cleared wine can be bottled. Another method of removal is to reduce the temperature of the wine to approximately 0 °C and seed it with finely ground tartrates, followed by a vigorous stirring. The seeds then attract further crystals to them and the entire process of removal takes just 24 hours or so. Vats being cold stabilised are shown in Fig. 11.5.

Fig. 11.5 Cold stabilisation

11.4 Adjustment of sulphur dioxide levels

As we have seen, sulphur dioxide is the winemaker's generally used antioxidant and disinfectant. However, only a percentage of the sulphur dioxide in the wine, the 'free' SO_2, acts in this way. The other proportion becomes 'bound' with the wine and is inactive. Before bottling, the free SO_2 levels should be adjusted to between 25 and 35 mg/l. Higher levels are needed for sweet wines to inhibit further fermentation of the sugars.

11.5 Bottling

A number of treatments may be carried out immediately prior to bottling to ensure the wine's final stability. These include pasteurisation or sterilisation. Four systems of treatment are available to the producer and use is dictated by the wine's specification.

- *Cold sterile filtration*: this process does not utilise heat. The wine is filtered through fine sheets or a membrane filter to remove all yeast cells. It is then aseptically bottled in a sterile environment. The method is particularly suitable for wines containing residual sugar and with modest alcohol levels. Such wines would otherwise risk re-fermentation in the bottle.
- *Thermotic bottling*: the wine is heated to 54 °C and bottled hot. At this temperature the yeasts are killed, and the alcohol becomes toxic to any bacteria.
- *Flash pasteurisation*: the wine is heated to 95 °C for one or two minutes, then rapidly cooled and bottled cold.
- *Tunnel pasteurisation*: the wine is bottled cold, and then passed through a heat tunnel where the sealed bottles are sprayed with hot water to raise the temperature to 82 °C for 15 to 20 minutes.

11.6 Closures

Corks are the traditional way of closing wine bottles. Cork is a natural product which comes from the bark of a type of evergreen oak tree, *Quercus suber*. Portugal produces over half the world's cork. A very important quality of cork is its elasticity, so if it is compressed it will try to regain its original size. Hence, when used as a closure for wine bottles, it will provide a tight seal so long as it is kept moist. Wine bottles should therefore be kept lying on their sides to prevent the cork from drying out. However, because corks are a natural product, the quality is variable and sometimes a taint can be imparted to the wine, as discussed in Chapter 17.

Synthetic materials have been developed as an alternative to natural cork. Plastic stoppers have now been on the market for some years, but are still proving controversial. A common consumer complaint is of the damage that plastic stoppers can do to corkscrews. Aesthetically, there can be no confusion that these are not natural corks, and indeed some producers choose vivid colours for the stoppers to complement the labelling. Another substitute for natural cork, which has been marketed for some years, is often described as a 'look-a-like' cork product. At a distance, this 'artificial cork' bears more than a passing resemblance to natural cork. Manufactured from a closed-cell polymer mixture, it is claimed that the closure has similar qualities to natural cork, including a minute porosity to gases and water vapour, enabling the wine to 'breathe'.

The metal screw cap (Stelvin®) closure is regarded by many as the best means of excluding oxygen, and a rapidly increasing number of producers are now turning to this, particularly for white wines. The choice of type of closure may well depend upon the buyer's specification, particularly when the buyer is a major supermarket group.

Detailed Processes of Red and White Wine Making

So far, we have studied the basics of red and white wine making. Now we will consider in more detail some of the equipment that may be used and processes that may be undertaken.

12.1 Wine presses and pressing

We recall that in red winemaking the press is normally used near the end of the fermentation to extract juice from the grape skins, and that in white winemaking the press is used before fermentation. The grapes may initially be gently bruised or crushed before pressing. Alternatively, whole clusters of grapes may go straight to the press. There are many different types of press, and the type used will depend on a number of factors, as discussed below. A wine press may be categorised as either a **continuous press** or a **batch press**.

12.1.1 Continuous press

There are many variations of the system of continuous pressing, but all work on the principle that grapes are fed into one end, carried along the length of the press by an Archimedes screw, and are subjected to ever increasing pressure as they continue their journey. Must can be collected at various points, with better juice coming from the earlier collection points. However, none of the must will be of very high quality. As long as grapes are being fed in, the machine will work. Continuous presses are mostly used in large 'industrial' wineries, having the advantage of efficiency in processing copious quantities of fruit.

12.1.2 Batch press

This category can be divided into three basic types: the horizontal plate press; the horizontal pneumatic press; and the vertical basket press.

12.1.3 Horizontal plate press

This type is very often referred to as a 'Vaslin press', being the name of the original company of manufacture, just as we often call a vacuum cleaner a Hoover. (NB: The present day company Vaslin–Bucher makes a number of different types of press.) A horizontal press consists of a perforated cylinder containing a system of plates, hoops and chains. Grapes are loaded into the press through a hatch which is then closed, and the press rotates, tumbling the grapes. At this stage, high quality free run juice is extracted. Then two metal plates move from each end along a screw towards the centre of the press, squeezing the grapes between them and releasing good quality juice. The cylinder then rotates in the opposite direction, and the plates are returned to the ends of the press. For the next pressing, the press first rotates, and the 'cake' of pressed grapes is broken up. The plates return to the centre forcing more juice, but of a lower quality, from the grapes. The press may operate three or more times in this way. In each grape berry, the best quality juice is in the area called *Zone 2* – this is neither close to the skin, nor the pip, and it is this juice that is released in the first pressing. Plate presses are not the most expensive, and are particularly popular with smaller properties making red wines. An illustration of a horizontal plate press is shown as Fig. 12.1.

12.1.4 Horizontal pneumatic press

This type of press is sometimes referred to as a Willmes or Bucher press, again after the names of the original companies of manufacture. Pressing in a pneumatic press is a relatively gentle, low pressure, operation. An illustration of a pneumatic press is shown as Fig. 12.2.

The press rotates, and a central rubber bag, bladder or sheet is inflated by means of compressed air. The grapes are pressed against the entire slotted cylinder – a much larger surface area than in the plate press. Again, several operations will take place to extract all

Fig. 12.1 Horizontal plate press

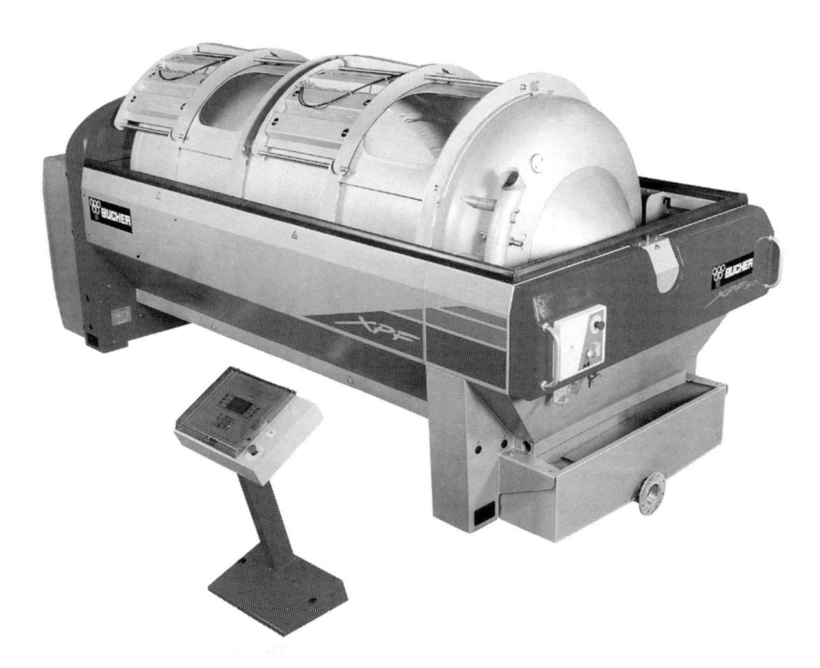

Fig. 12.2 Pneumatic press

Fig. 12.3 Inert gas injected into tank press

the usable juice, with the finest must being from the first pressing. A tank press (Fig. 12.3) is a fully enclosed pneumatic press, which can be flushed before use with an inert gas such as nitrogen or carbon dioxide, thus helping to eliminate the risk of oxidation of the must.

Pneumatic presses are particularly suitable for white wine vinification, the gentle process helping to retain aromatics from the grapes. Nowadays, the press or even a whole bank of presses may be computer controlled, adding to efficiency (and capital costs). Pneumatic presses are expensive – up to £40 000 depending upon capacity – and they are only used for a few days in the winemaker's year. Thus, the owners of small wineries are very keen to proudly display their recent purchases to any visiting wine lover who appears at all interested.

12.1.5 Vertical basket press

This is the traditional press, as developed by the monks of the Middle Ages. A wooden plate is lowered onto grapes contained in a basket usually made up of wooden staves. There are several variations on the basic theme, including deep or shallow baskets, the latter still being extensively used in the Champagne region of northern France. Basket presses are very gentle.

The press can be used for whole clusters of grapes in white wine making, or for the remaining solids for reds. The larger the diameter of the press, the more quickly the juice will run away from the bed of grape skins. The stalks help provide drainage channels. However, the draining juice is exposed to the air all along its route, and it is important to process it without delay in order to limit oxidation. A basket press is shown as Fig. 12.4.

The use of basket presses is a particularly slow process, both in pressing and cleaning. With horizontal presses, once the operation is complete, the cylinder is inverted and the cake of dry skins is emptied through a hatch so the press can be cleaned. Vertical presses require the remaining pips and skins to be removed manually and the basket must be cleaned thoroughly. Nevertheless, they are gaining in popularity again, particularly with many New World winemakers. One

Fig. 12.4 Vertical basket press

specialty use is for a particular style of red wine. Because of the nature of the pressing, the grape solids remain reasonably static, resulting in a press-wine that may be less bitter than would be obtained from a plate press. This can allow the spicy flavours of certain varieties to prevail. Much technology in winemaking seems to be going full circle – many quality winemakers are now also replacing their stainless steel vats with wood!

12.2 Use of gases to prevent spoilage

At many stages during the winemaking process, inert gases may be used to prevent oxidation and other spoilage. We have already considered their use at the pressing stage. The gases commonly used are nitrogen (N_2), carbon dioxide and, increasingly, argon (Ar). Carbon dioxide is highly soluble in wine, and its use can give the finished wine a 'prickle' on the tongue – particularly undesirable in red wine making. Nitrogen, however, having a lower molecular weight, does diffuse quickly. Accordingly, mixtures of two of the gases have advantages.

Before fermentation, vats may be sparged; partially filled vats may be blanketed; and at racking and bottling times empty vessels may also be flushed with the gas. Depending upon the gas used, considerable amounts are required to flush all the air from the container. The average amount of gas used in producing a (9 litre) case of Australian wine is 100 litres – for high quality wine this rises to 600 litres!

12.3 Natural or cultured yeasts

A decision is made whether to use the natural yeasts on the grape skins or to inhibit these and work with cultured yeasts. Generally speaking, most Old World producers favour the use of natural yeasts, particularly for red wines, believing that they are a part of the 'terroir' of the vineyard, and contribute to the individual character of the wines.

In the New World, most producers consider natural yeasts far too risky to use. Control is the aim, and cultured yeasts can be selected for their individual characteristics, including flavours, ability to perform at various temperatures and formation of lees.

12.4 Destemming

There are winemakers who believe that stems can perform a valuable function. For example, in the Rhône valley, stems are often retained in the vinification process. They help keep the cap of skins on top of the fermenting must less compact, and do add firm tannins to the wine and suppress any tendency for overt jammy fruitiness.

12.5 Fermenting sugar-rich musts to dryness

In hot climates, grapes are sometimes so sugar rich that difficulties may arise in achieving fermentation to dryness. The problem may be overcome by the (normally unmentionable) addition of water to the must or in the early stages of fermentation. Winemakers may refer to 'flushing out the picking bins very well', or using the help of 'the black snake' or 'the white horse'.

12.6 Colour extraction, concentration and tannin balance

There are several methods that may be used to extract colour, concentrate flavours and achieve tannin balance.

12.6.1 Must concentrators and reverse osmosis

Grapes may be picked that are wet from rainwater or that have not reached complete ripeness. The inclusion of rainwater at pressing can cause dilution of juice solids (sugars and other components). When under-ripe grapes are pressed the sugar-in-juice level may be lower than required to make a good wine with the right level of alcohol and the required balance. In both instances the solution may be to remove water.

Must concentrators use vacuum evaporation to remove water, thereby increasing the level of sugars, colour and flavour components and tannins in the must. A must concentrator manufactured by REDA is shown in Fig. 12.5. An alternative technique is to use reverse osmosis. This is a membrane separation process in which a microporous membrane acts as a molecular sieve, allowing water molecules to pass through, but trapping larger molecules (sugars, etc.) which are then concentrated. Reverse osmosis can yield must capable of creating full bodied, complex and higher quality wines. It can also be used to reduce the alcohol level in over concentrated fermented must.

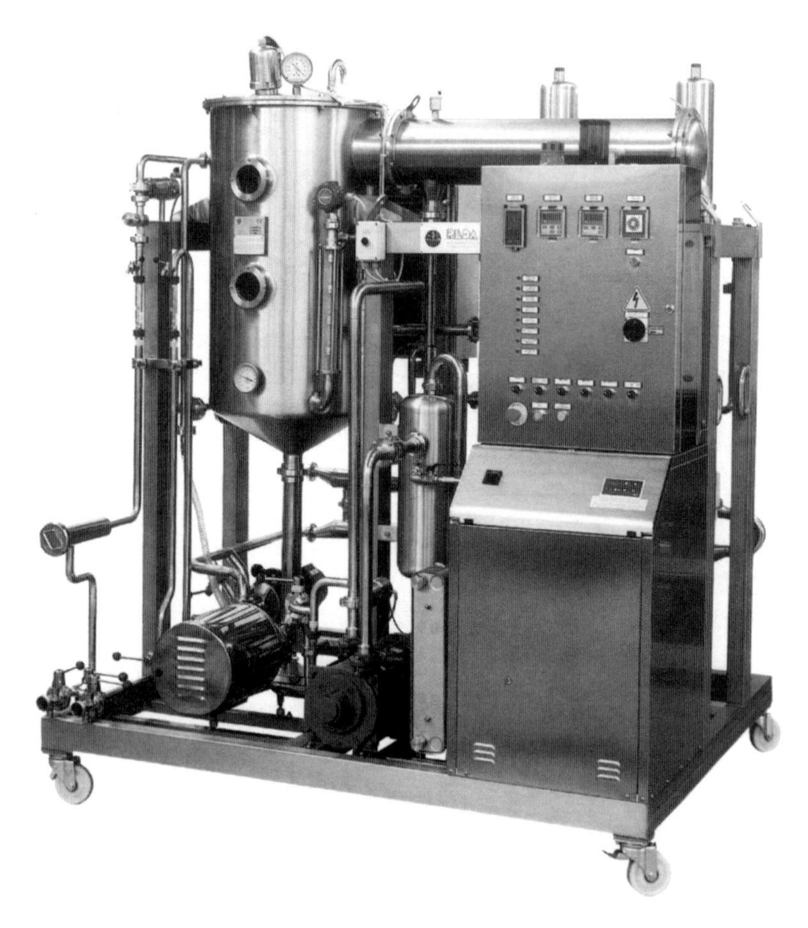

Fig. 12.5 Reverse osmosis equipment

12.6.2 *Cold soaking (pre-fermentation maceration)*

Many producers hold the must at a low temperature and carry out a pre-fermentation maceration on the grape skins. This method was introduced in the Burgundy region of France by Guy Accad, an oenologist who from the late 1970s has had much influence upon the winemaking at many top estates. Accad showed that soaking the skins in the pre-fermentation (watery) juice results in fresher, cleaner wines with greater finesse than when the skins are macerated in the post-fermentation (alcoholic) liquid.

12.6.3 *Pump overs*

Crucial to the quality and style of **red** wines is the extraction of colour and flavour giving compounds from the grape skins. There are several methods for achieving this which we will consider further. **Remontage** (pumping over) is very widely practised. In its simplest form, fermenting juice is drawn from near the bottom of the vat and pumped to the top, where it is sprayed over the cap of skins, pushed to the surface of the wine by the rising carbon dioxide gas. The operation may be carried out manually, including physically moving the hose to spray and soak all the skins. Vats may be fitted with timed automatic systems that pump the juice up a fixed pipe at the side of the vat; the juice then falls onto a rotating diffuser, which gives a more gentle spray onto the cap. When additional aeration is required, as the juice is drawn from the vat it is allowed to splash into a large vessel from where it will be pumped up, as is shown in Fig. 12.6. The number, timing and method of pump overs will impact upon the style and quality of the finished wine.

12.6.4 *Rack and return (délestage)*

Historically, even relatively inexpensive red wines would often be laid down for a few years to soften the tannins; for the classic reds an extended period of bottle maturation was essential. Nowadays, most wine is consumed within a few days of purchase, and often within a few months of bottling. Accordingly, techniques to extract

Fig. 12.6 Aeration during pump over

maximum flavours, but with soft tannins, are required. One procedure now widely undertaken is **rack and return**, also known as **délestage**. One of the problems with pumping over is that there can be a channelling effect – the descending juice flows down the same channels in the wine, beating out hard tannins from the adjacent skins. With the rack and return technique, most of the contents of the vat are racked to another vat, leaving behind the cap of grape skins, which loosens and falls to the bottom, breaking up somewhat as it does so. The fermenting juice is then returned, and the cap rises to the top, leaching much colour in the process. It should be noted that the total tannins extracted using délestage may be greater than with pump overs, the advantage being that less hard tannins are leached. Most winemakers feel that délestage should not be undertaken at a late stage in the fermentation process, as it can result in a greater astringency.

12.6.5 Rotary vinifiers

These are large, horizontal, cylindrical vessels that rotate during the fermentation process, thereby tumbling the grapes and promoting contact between the skins and juice. They eliminate the need for punching down or pumping over, and quicken fermentation and maceration. Such processing is particularly suitable when the aim is a soft, easy drinking red of medium body.

12.6.6 Thermo-vinification – heat extraction

The name thermo-vinification is often used for the process of extracting colour for red wines by heating crushed grapes for 20 or 30 minutes to a temperature of between 60° and 82 °C. This breaks down the vacuoles of the skin cells that contain anthocyanins and tannins, rapidly releasing the pigments into the juice. After pressing and cooling of the juice, the fermentation takes place off the skins, in the manner of a white wine. Wines made using this method often have a vibrant purple colour. They can sometimes taste rather 'cooked' and lack definition, so the process is not used for fine wines.

12.6.7 Whole grape fermentation, carbonic and semi-carbonic maceration

Most Beaujolais and many other light bodied red wines are made by a distinctive method, which we shall call whole grape fermentation. Whole bunches of grapes are tipped into a vat that has been sparged with carbon dioxide. To a degree, the grapes in the bottom of the vat are pressed by the weight of those above. Within the berries that remain whole, an enzyme is secreted resulting in a fermentation up to approximately 3% alcohol, without the action of yeasts, and this contributes to the jammy style of the wines. Thereafter the yeast enzymes take over to complete the fermentation. The fermenting wine is allowed to macerate or soak on the grapes for between 4 and 10 days, depending on the style of wine required. The activities taking place in the different levels of the vat are very complex.

Fig. 12.7 Staves before insertion in vats

12.6.8 *Fixing colour*

Steps may be taken to ensure that a red wine will hold its colour. The use of inner staves in a vat of wine is one way of achieving this, the tannins released by the wood binding the colour to the wine. The process is often used in New World countries – it is not generally permitted for wines made in the European Union. Fig. 12.7 shows inner staves before installation in a vat.

12.6.9 *Post-fermentation maceration*

Historically, after fermentation red wine often had a very lengthy period macerating on the skins. During this time tannins would be absorbed, which would help to fix the colour and gives the wines structure, but the resulting wine would often be very firm in the mouth. During the past 20 years, many producers have shortened this period, especially for wines that are not destined for lengthy bottle ageing.

12.7 Macro-, micro- and hyper-oxygenation

As we have seen, oxygen can be both friend and enemy to wine. It is now generally accepted that during fermentation there is little chance of the wine oxidising, and indeed oxygen is necessary to maintain healthy yeast colonies. After fermentation, wines may benefit from absorbing tiny amounts of oxygen, for this will help polymerise the polyphenols and help fix the colour of the wine. There are three methods by which wine may be deliberately oxygenated, depending upon the purpose.

12.7.1 Hyper-oxygenation

This is adding large amounts of oxygen to white must pre-fermentation. The benefits are a reduction in **volatile acidity** and **acetaldehyde**.

12.7.2 Macro-oxygenation

This is adding large amounts of oxygen during the fermentation. It may be undertaken as 'rack and splash', in which the wine is aerated when returned to the vat, or can be incorporated during pump overs. Alternatively a machine known as a 'cliquer' may be used. Macro-oxygenation can also be used to try and restart stuck fermentations, but it is commonly undertaken to soften harsh tannins.

12.7.3 Micro-oxygenation

Micro-oxygenation is the continuous addition of tiny amounts of oxygen to wine in order to improve aromas, structure and texture. It may be carried out before or after the malolactic fermentation, with the former perhaps giving the greatest benefits. The process replicates the absorption of oxygen through the wood that occurs naturally when wine is matured in barrels. The process must be carefully controlled. It is essential to know the amount of dissolved oxygen in the wine at the outset, and the rate and amount of the addition, for there should be no accumulation of dissolved oxygen in the wine during the process.

Micro-oxygenation can reduce undesirable herbaceous or sulphur characters. It also helps to stabilise the wine.

12.8 Removal of excess alcohol

Reverse osmosis is one relatively inexpensive method of removing excess alcohol from wines. An alternative method is to draw off a proportion of an over alcoholic wine and send this to a specialist facility that utilises a spinning cone column (SCC). This is a distillation column through which wine is passed twice. The first distillation extracts the volatile aromas and the second removes the alcohol. The aromas are then returned to the 'wine', which is returned to the winery and blended back into the bulk. *ConeTech* at Santa Rosa, California, operate the world's largest SCC plant.

Barrel Maturation and Oak Treatments

There is no doubt that oak maturation is very fashionable. The back labels of countless bottles contain such phrases as 'subtle hints of oak' or 'wonderful oaky superstructure'. Visions immediately come to mind of row upon row of barrels of wine gently maturing in cool, dark cellars. However, inexpensive wines are very unlikely to be matured in oak barrels, for economic reasons. The cost of a new French oak barrique, which hold 225 litres (300 bottles) of wine, is in the region of £380–£500. When barrels are new, they impart many oak products to the wine contained therein, including vanillin, lignin and tannin. When the barrels are used again (second fill), the amount of oak derived by the wine is reduced. After the barrels have been filled four or five times, they become merely storage containers that allow the wine to mature as a result of a very slow, controlled, oxygenation. Achieving a fine balance of oak aromas and obtaining the right level of oak influence is the art of a skilled winemaker. Over-oaked wines may well smell of burnt toast, fatty butter and marmalade, and taste pseudo-sweet, with all the fruit of the wine enveloped in a sea of coarse wood products. In other words, oak is a tool for adding complexity, and should not be the predominant feature in a wine.

13.1 The influence of the barrel

The impact of the barrel upon the style and quality of wine will depend on a number of factors, including size of barrel; type and origin of oak (or other wood); manufacturing techniques, including toasting; amount of time spent in barrel; and where the barrels are stored.

13.1.1 Size of the barrel

The smaller the barrel the greater the surface to volume (wine) ratio and thus the greater the influence of the oak. A 225 litre barrique will give 15% more oak products to a wine than a 300 litre barrel.

13.1.2 Type and origin of oak (or any other wood)

American oak is coarser grained than French oak, and the wood is usually sawn, whilst French oak barrels are made from wood that is split along the natural grain of the wood. Within France, there is a distinctive difference in the oak from the various forests: Allier, Bertagne, Limousin, Nevers, Tronçais, Vosges, etc.

13.1.3 Manufacturing techniques, including toasting

The techniques used in the coopering of barrels vary from country to country. Barrels made of American oak but coopered in France are very different from, and often preferred to, those coopered in the USA. In the latter stages of manufacture, barrels will be 'toasted' to a greater or lesser degree, i.e. the insides, or maybe just the 'heads' (ends), will be charred over a small wooden fire. The type of wood used to fuel the fire will have an impact upon the toasting, and consequently the wine stored in the barrel.

13.1.4 Amount of time spent in barrel

Historically wines were often kept in barrel until they were required for shipping – in some instances this might be as long as 10 years. Nowadays it is exceptional for any wine to be kept for more than three years in wood. The longer the period, the more oak products will be absorbed, the more oxygenation, possibly resulting in oxidation. The barrels will need topping-up shortly after filling and in the early stages of storage. Initially barrels will be stored with the bunghole on top, with the bung loosely inserted or with a fermentation lock in place. Later the barrels may be stored with the sealed bung hole on the side, the contact with the wine ensuring the bung remains expanded in the hole.

During the first year in barrel, the wines may be racked four times, and during the second year, they may be racked twice.

13.1.5 Where barrels are stored

Barrel maturation is an interplay between the wine, the wood and the atmosphere in which the barrels are stored. Wine evaporates through the wood and is replaced by oxygen, and oak products are absorbed. The process is temperature and humidity dependent. For example, in a warm, dry atmosphere, the volume of wine that evaporates is greater than in a cool, damp atmosphere. However, in a cool, damp atmosphere, a greater volume of alcohol evaporates. Thus a wine from an individual property that is barrel matured in an underground cellar may be subtly different to a previously identical wine that is matured in a warehouse above the ground.

13.2 Oak treatments

As we have seen, maturation of wines in new oak barrels is an expensive process. It is not just the high cost of the barrels, but also the extra labour in racking and working with small individual batches. Thus inexpensive wines from New World countries are likely to have been matured in vats, and if 'oaked' this usually consists of oak chips, the waste from cooperage, being soaked in the wine. There is not, of course, the interplay between the wine, the wood and the atmosphere, and chipped wines often exhibit a sweet, vanilla sappiness. The level and style of oakiness depends upon the amount and size of the chips, the source of the oak, how the chips have been toasted and, of course, the length of time of the maceration.

More subtle oaking can be achieved by the insertion of oak rods, planks or oak staves into vats of wine. The staves may be new and fire toasted, or may have been 'reconditioned' from broken down old barrels. They can be built into a tower, or inserted into the tank as a module, rather like an inner lining to the sides of the tank. This system allows for a high level of extraction together with total control yet with, of course, considerable cost savings over barrel maturation.

Even if bottle labels refer to 'barrel maturation', this does not necessarily mean that all the oakiness has come from the actual barrels. Oak chips or rods may be suspended in perforated sleeves through the bunghole, and moved from barrel to barrel, as required. Alternatively, the barrels may have undergone a 'renewal system' – the insertion of new staves inside the barrel, or the building of a grid of oak sticks.

Producing an inexpensive wine with subtle oak tones is certainly an art, and many different methods may be combined. For example, Jürgen Hofmann, chief winemaker at the award winning Reh-Kendermann winery, uses a blend of wines that are 80% chip treated, 10% stave treated and 10% barrel matured to produce a delightfully balanced Romanian Merlot. However, it is perhaps sad that many consumers, and winemakers, have become too obsessed by oak flavours, but are not prepared to pay the price for genuine barrel maturation.

Making Other Types of Still Wine

So far, we have discussed the procedures for making red and dry white still wines. Now we will look very briefly at ways in which other styles of wines are made.

14.1 Medium-sweet and sweet wines

Generally speaking, following natural fermentation, most or nearly all of the sugars will have been consumed by the yeast. The European Union has legal definitions for the description of the terms sweetness and dryness. For example, wines containing less than 4 grams of residual sugar per litre are described as 'dry'. At the other end of the scale, a wine described as sweet must have not less than 45 grams per litre.

If medium-sweet or sweet wines are required, there are several ways they can be produced. The method used will depend upon the ripeness of the fruit available, and the style, quality and cost of the wine aimed for.

Fermentation can be stopped deliberately while a quantity of sugar remains in the wine. Historically, winemakers would use a heavy dose of sulphur dioxide to kill off the yeasts. Nowadays, the quantity used is strictly controlled by law. One method of stopping fermentation is by cooling the must so that the yeast becomes dormant and inactive. The wine will then be filtered to remove the yeast cells. Another way is for the alcohol level to be raised by the addition of grape spirit (as in Port winemaking). Depending on the yeast strain, 15% alcohol is about the maximum level to which fermentation is possible.

Particular techniques can be used for making medium-sweet or sweet wines.

14.1.1 Medium-sweet wines

These can be made by the addition of a small amount (10–15%) of sterile unfermented grape must after the wine has finished fermentation. This is a practice commonly followed in Germany for inexpensive wines, where the juice added is known as **Süssreserve**. This helps reduce any aggressive acidity (and the alcoholic degree), and helps maintain fresh grapey flavours.

14.1.2 Sweet wines

In certain vineyards and/or in exceptional vintages, grapes can ripen to such an extent as to have very high sugar levels. As we have seen, the concentration of sugars may have been increased by *Botrytis cinerea* in the form of noble rot. Just as yeast struggles to work at low temperatures, high sugar levels can reduce the yeast's alcohol tolerance. Thus fermentation may stop before all the sugar has been converted. High alcohol levels and high residual sugars can be achieved, as for example in Sauternes, from Bordeaux, which has around 13–14% alcohol and sugar levels in excess of 150 grams per litre. However, some very sweet German wines, such as Trockenbeerenauslese, may only ferment to around 8% alcohol.

Let us consider the making of a typical Sauternes: Château Bastor Lamontagne. The grapes are successively picked and pressed in a horizontal press. The first pressing gives must with the most aromatics and the third, last pressing gives juice with the highest sugars. The pressings are blended according to the winemaker's assessment. Natural yeasts are used, and even though the fermentation temperature is 20° to 22 °C, this will take from 3 to 6 weeks, depending upon the richness of the must. The relatively high fermentation temperatures help give the wine body and richness. All the fermentations are started in 125 hectolitre vats. Some of the fermenting wines are then transferred to two- to three-year-old barriques to continue; others remain in vats. During the fermentation it may be necessary to heat the barrel cellar. When the winemaker considers that the wines have the correct balance of alcohol and sugar the fermentation is stopped, in the case of the vat component by chilling these to 4° or 5 °C and adding sulphur dioxide. Maturation takes place in both vat and barrel for 14 months or so, with the components being blended shortly before bottling.

Traditionally, in some parts of the world, grapes are dried or 'raisined' as a means of concentrating sugar levels. This may take place on grass mats in the fields under the sun. An alternative practice is carried out in some areas. For example, in Veneto in Italy, Recioto Della Valpolicella is made from grapes that have been laid on trays or hung from rafters in ventilated lofts. During the 4 months or so that the grapes are stored this way they lose up to 50% of their water content, thus resulting in a concentration of sugars. With such high sugar levels, fermentation is a very slow process and can take months.

14.2 Rosé wines

Rosé wines can vary in colour from the palest of pinks to light red. 'Blush' is another term for the palest rosés that became popular in California in the 1980s. The Spanish use the term 'rosado'. 'Clairet' is a style of deep rosé from Bordeaux. The sweetness levels of rosés can vary from dry, to off dry, to medium-sweet. Alcohol levels vary considerably too, depending on grape ripeness and vinification techniques.

There are several methods used for producing these wines:

- *Blending*: the blending of red and white wine together is now permitted in the European Union. Historically this method could not be used, except in the Champagne region where it has long been the main method for the production of rosé Champagne.
- *Skin contact*: this style of rosé is made from red grapes or white grapes with heavily pigmented skins, such as Pinot Gris, using the white grape method of winemaking. Some grapes are crushed and strained; others are pressed. The free run juice together with the lightly pressed juice is then fermented without the skins. Pigments from the skins give some degree of colour to the juice. Quite pale rosés often result from this method and are sometimes labelled as 'vin gris' (grey wine). Many rosé wines from the Loire are made in this way, as are California's 'Blush' wines. If extra colour is required a little red wine may be blended in.
- *Saignée*: this French term means 'bleeding'. The wine may be made by a short (6 to 12 hours) fermentation on the grape skins; then the juice is drained. It is also possible to produce a rosé

'by-product', by draining part of a vat for a rosé and leaving behind a great concentration of skins to produce a full-bodied red wine. Alternatively, red grapes can be chilled and lightly crushed just sufficiently to release their juice. They are then allowed to macerate for between 12 and 24 hours, during which time the colour pigments and flavourings are allowed to 'bleed' into the juice. The juice is then drained off and fermentation begins. This method can produce quite good depth of colour, especially if the grapes used have deep coloured skins, e.g. Cabernet Sauvignon. This style of rosé is likely to be firmer, often showing some tannins.

14.3 Liqueur (fortified) wines

This category of wines has higher alcohol levels than the light wines previously discussed. Indeed, they are often referred to as 'fortified', since a quantity of grape spirit is added to 'fortify' them. The resulting alcohol levels in these wines can vary between 15% and 22% ABV. Historically, brandy was added to light wine in barrels as a means of preservation and stabilisation, especially on long sea voyages. Detailing the production of liqueur wines is an extensive subject, but we will consider the topic very briefly here.

One method of production of liqueur wines involves fermentation to dryness before the addition of grape spirit. The other basic method involves the addition of grape spirit during the fermentation process, taking the alcoholic degree above that at which yeasts will work, resulting in naturally sweet wines.

The most well-known examples of these wines are Sherry and Port. A crucial difference in the method of making the wines is the timing of the addition of the grape spirit. Sherry is initially fermented to dryness; Port has the spirit added during the fermentation.

14.3.1 Sherry production

Sherry is the liqueur wine produced in Andalucia in south-western Spain, and the processes now detailed are those that apply to wines made in that area. Once fermentation has ceased, leaving the wines fully dry, they are initially classified as either 'Finos' or 'Olorosos'.

Wines are transferred into barrels known as 'butts'. The butts are not completely filled, leaving a small air gap over the top of the wine. A feature of Sherry production is a naturally occurring yeast (*Saccharomyces beticus*) which may grow on the surface of wines in some barrels or vats, creating a yeasty crust known as 'flor'. Flor tends to grow thickest in cooler coastal areas. After fermentation, wines destined for the Fino style are lightly fortified to 15%, which is the ideal alcoholic degree at which flor develops. Flor also requires oxygen and nutrients (which are naturally present in the wine) to live. The crust of flor protects the wines from oxidation – the ageing of the wines is microbiological. If flor dies back, microbiological ageing will cease and chemical changes will take place. Oxidation will occur and the wines will naturally darken. These wines may be designated as 'Amontillados'.

Wines of the heavier Oloroso styles will have been fortified up to 17% or 18% after fermentation. Flor will not grow at these alcoholic degrees and thus the wines will age in an oxidative manner. They may ultimately be sweetened, as for 'cream sherries'. Sherry wines will then mature in a 'solera system', which is a means of blending young wines with old to ensure consistency of style and quality. In order to be legally sold as sherry, the wines have to be aged for at least three years in the solera system.

14.3.2 Port production

The Upper Douro River in Northern Portugal is home to the vineyards producing grapes for Port wine. A most important feature of Port making is the extraction of colour and tannin from the grape skins. Traditionally, this was done by treading the grapes by foot, still considered by some to be the gentlest way. More mechanised methods are now generally used. During the early stages of fermentation, it is critical that the skins are kept in contact with the juice to continue the extraction of colour and tannins. The fermentation process is then interrupted, when there is somewhere between 6% and 9% alcohol, according to the house style. The juice is drained off and fortified to between 18% and 20% ABV. Yeasts will not work at this level of alcohol and so fermentation is stopped. There are various styles of Port, including Ruby, Tawny, Late Bottled Vintage and Vintage. Each of these styles has its own particular length and style of maturation. Some styles involve decades of barrel maturation; others are bottled early and need decades in bottle.

14.3.3 Other well-known liqueur wines

Liqueur wines are made in many wine producing countries using an array of production methods. We will very briefly consider a few of these.

- **Madeira**: sweeter styles are vinified in a similar manner to Port, whilst drier styles are fortified after fermentation.
- **Marsala**: from Sicily, a fully fermented wine which is fortified with grape spirit and usually sweetened with boiled down must (known as mosto cotto).
- **Málaga**: this classic wine from Spain is vinified in a similar manner to Port, but is often aged in the solera system as for Sherry.
- **Vins Doux Naturels**: a range of 'naturally sweet wines' produced in many parts of southern France. These are mainly white, made from Muscat grapes, although a few reds from south-west France are made from Grenache. They are naturally sweet, having had the fermentations arrested.
- **Commandaria**: produced in Cyprus. After picking, the grapes are 'raisined' in the sun. Fermentation usually stops before all the sugar has been consumed, leaving a naturally sweet wine. Following this, grape spirit is added to give a final alcohol of 20%.
- **Moscatel de Sétubal**: produced in the Sétubal Peninsula in Portugal in a similar manner to Port.

Sparkling Wines

The bubbles in all fizzy drinks, including sparkling wines, are carbon dioxide. In the case of most drinks other than wine, the CO_2 is injected into the liquid before bottling. Whilst it is true that some wines (e.g. Lambrusco) are given a spritz by this injection method, the sparkle in all quality sparkling wines is a result of retaining in the wine the natural CO_2 produced by fermentation. There are two basic methods by which this may be achieved: **fermentation in sealed tank** and **second fermentation in bottle**.

15.1 Fermentation in sealed tank

To retain carbon dioxide in the wine from the fermentation, this may take place in a sealed tank. There are two variations according to the style of wine required: the **continuous method**, one fermentation as in the case of Asti DOCG; and the **Charmat method**, a second fermentation as in the case of Veuve du Vernay.

15.2 Second fermentation in bottle

A second fermentation may be undertaken in a bottle. There are two variations: the **traditional method**, a second fermentation in the bottle in which the wine is sold, as in the case of Champagne; and the **transfer method**, a second fermentation in a **cellar** bottle (not the bottle in which the wine is sold), as in the case of many Australian sparkling wines.

We know from our study of winemaking that fermentation generally depends on the yeast cells in suspension. When the fermentation is finished the yeast will fall to the bottom of the vessel as lees, unless agitation takes place to keep it in suspension.

Before the making of a sparkling wine is finished, the yeast or yeast sediment must be removed. In the **Charmat method**, also known as **cuve close**, this is simply a matter of filtering out the yeast and bottling the wine under pressure. In the **transfer method** the bottles are decanted into a vat under pressure, and the wine is filtered and rebottled, still under pressure. However, the **traditional method** requires a much more complicated system of sediment removal, which was invented in the Champagne region and is often referred to by the French terms of **rémuage** and **dégorgement**.

The traditional method has, by law, to be used for making of Champagne. All other French AC sparkling wines have also to be made by this method (or variations upon it, e.g. **méthode gaillacoise**), and it is the method used for the finest quality sparkling wines produced elsewhere in the world.

15.3 The traditional method

We will consider the sequence of events undertaken to produce a traditional method wine as practised in the Champagne region.

15.3.1 Pressing

This may take place in a horizontal press (either pneumatic or plate) or in a traditional shallow vertical 'cocquard' basket press. The capacity of a basket press (and a common size for horizontal presses) is 4000 kilograms (4 metric tonnes) of grapes. There is a legal limit in the Champagne region on the amount of juice that may be used for Champagne – 2500 litres per press charge of 4000 kilograms. This equates to 100 litres of juice per 160 kilograms of grapes (1 litre per 1.6 kilograms which, allowing for rackings, means that nearly 1.5 kilograms of grapes is needed for each bottle). The first 80% of the juice extracted (2000 litres) is called the **cuvée**; the last 20% (500 litres) is called the **tailles** (tails) and is likely to be included in the blend used for cheaper Champagnes.

When pressing the grapes for Champagne it is important not to extract colour from the two red grape varieties (Pinot Noir, and its relative Pinot Meunier) that form the backbone of most Champagnes.

15.3.2 *Débourbage*

The juice is pumped into vats, which are chilled, and sulphur dioxide (and perhaps some bentonite) may be added to assist solids to settle. It is particularly important to ferment clear juice for sparkling wines.

15.3.3 *First fermentation*

In most cases this takes place in vats – stainless steel, cement or concrete. A few Champagne houses ferment in barrel, e.g. Krug, Alfred Gratien and the vintage wines of Bollinger. The fermentation is undertaken at relatively warm temperatures, 18° to 20 °C, as a warm fermentation reduces the unwanted aromatics. The first fermentation (to a level of 10.5% or 11%) will take approximately 2 weeks, after which most wines undergo a malolactic fermentation, although some houses block this. The wines will be racked and are now called **vins clairs**.

15.3.4 *Assemblage*

This is the blending of the vins clairs. Champagne is usually made from a blend of varieties: Pinot Meunier, Pinot Noir and Chardonnay, and the grapes grown in different parts of the region will have been vinified separately. Except in the case of Vintage Champagnes, **reserve** wines (vins clairs from previous harvests) may also be added: thus 30 or 40 different base wines may go into the blending vats.

15.3.5 *Addition of liqueur de tirage*

Tirage is the French word for bottling. Selected cultured yeasts and between 22 and 24 grams per litre of cane sugar are added to the vats of blended wine before bottling in the heavy sparkling wine bottles. Nowadays these are usually sealed with crown corks containing a small plastic cup (**bidoule**), although the finest wines and those from the smaller producers are often sealed with a traditional cork (**liège**) and staple (**agrafe**).

15.3.6 Second fermentation

The bottles are taken to the cool cellars where they are laid horizontally for the second fermentation. The yeasts work on the added sugar, making an extra 1.3% to 1.5% alcohol, and carbon dioxide which dissolves in the wine. The gas increases the pressure in the bottle to 75 to 90 lb per square inch (5 to 6 bar). The process may take between 14 days and three months, dependent upon temperature. The longer and slower the fermentation, the better the resultant quality.

15.3.7 Maturation

The bottles will be stored horizontally for the wine to mature, including the development of **autolitic** (yeasty) character. For Champagne the minimum legal maturation period is 15 months, and for Vintage Champagnes three years. From time to time the bottles may undergo **poignettage** – shaking and restacking – to prevent the yeasty sediment from sticking to the glass.

15.3.8 Rémuage

This is the process of riddling the sediment into the necks of the bottles. Traditionally the bottles are placed into wooden easels called **pupitres**, each side of which contains 60 slanted holes as shown in Fig. 15.1. The bottles are inserted almost horizontally, neck first. Every day for between six weeks and three months a skilled cellar worker (rémuer) will give the bottle a twist (one-eighth of a turn), a shake and move them so that they are slightly more vertical. At the end of the period of rémuage, the yeasty sediment will have been coaxed from the sides of the bottles into the necks. Using this system, the lighter sediment makes its way into the neck first, followed by the heavier.

All the large Champagne houses (and other wineries that make sparkling wines by the traditional method) now use an automated process of rémuage. The most commonly used system is that of **gyropalettes**, as shown in Fig. 15.2. These machines comprise a number of wooden containers, each holding 504 bottles. Controlled by a computer, the machine replicates the movements of the rémuer.

Fig. 15.1 Pupitres

Fig. 15.2 Gyropalettes

Some smaller houses have gyropalettes that have to be cranked by hand, and some use a machine that looks like an automated pupitre – **pupimatic**.

15.3.9 Stacking sur pointes

This is undertaken by some Champagne houses for all wines, and by others for the best wines only. The bottles are stacked almost vertically with the neck of one bottle into the base of another for a period of up to three months.

15.3.10 Dégorgement

This is the disgorging of the yeasty sediment. The usual system used today is **dégorgement à la glace**. The bottles, with the sediment in the necks, are transported (still inverted) and the necks are inserted into tubs containing a freezing mixture. After approximately 6 minutes, the yeast becomes encapsulated in a slightly slushy ice pellet. The bottles are removed, turned the right way up, and a machine removes the crown cork. The pressure of the carbon dioxide ejects the yeast sediment.

A few smaller houses still disgorge **a la volée**. The bottles are held pointing upwards at an angle of 30° and the cork or crown cork is removed manually and the sediment expelled into a hood.

15.3.11 Dosage (liqueur d'expedition)

The bottles are now topped up with a mixture of identical wine and a small amount of cane sugar, the precise amount varying according to the required style of wine.

15.3.12 Corking

The final cork is now inserted, and wired on. Labelling and the addition of the foil will always only be undertaken just before shipment, in order to avoid damage in the cellars.

Problems and Solutions

16.1 Vintages – style and quality

The expression 'vintage wine' is frequently misused, especially in the popular press, implying wines of quality and age. Any wine, unless it has been blended with wine from previous years, is 'vintage', however young or poor. However, it should be remembered that in the European Union it is illegal for simple Table Wines (without geographical origin) to state the vintage on the label.

'Was it a good year?' is one of the questions most commonly asked of wine merchants, especially when the customer is considering laying down the wine. Vintages vary in quantity, style and quality. We have already seen that frost at budding time, or cold and windy weather at flowering time, can result in a considerably reduced crop. Whilst plenty of summer sunshine is clearly desirable, if it gets too hot, above 40 °C (104 °F), the vines' metabolism can shut down, resulting in under-ripe fruit. A little rain in spring or early summer is also generally desirable, for if the vines are too stressed quality will also be reduced. Perhaps most desirable of all is a period of dry, warm sunny weather in the weeks before harvesting. Many otherwise great vintages have been wrecked by the autumn rains arriving whilst the fruit is still on the vine. Vintages vary in style: big rich wines in years when it has been hot, but when perhaps there has been less sunshine, the wines may be perfectly good, but lighter. Grape varieties ripen at different times; for example Merlot ripens up to three weeks before Cabernet Sauvignon. Thus in Bordeaux, for example, if the rains set in during late September the Merlots will probably already be fermenting, and wines dominated by this variety may be excellent, whilst the Cabernet-dominated wines will be much less good. On the other hand, if August and early September have

Table 16.1 The least good years for red Bordeaux

1960s	1970s	1980s	1990s
1963 – disastrous	1972 – very poor	1980 – fairly poor	1991 – fairly poor
1965 – disastrous	1974 – poor	1984 – fairly poor	1992 – fair
1968 – disastrous	1977 – very poor	1987 – fair	1993 – fair

been cool, with excessive rain, but then there is an Indian summer later in the month and into October, it is the wines majoring on Cabernet Sauvignon that may excel.

Many New World regions boast that they can ripen fruit in almost any year. However, it is true to say that even in less favourable climates there are fewer 'poor years' resulting in lower quantities of mediocre wines than was once the case. Let us consider the least good years of the past four decades for red Bordeaux, as shown in Table 16.1.

What has happened in the past 40 years that has resulted in the reduction of very poor years? Is this a consequence of global warming? There is little doubt that climate change has had a very modest impact on wine quality. However, the real advance has been improvements in the knowledge of growers about achieving higher quality fruit, and makers in processing less than perfect fruit, in years when the weather is unfavourable. We will consider just a few of the techniques used.

16.2 Coping with problems in the vineyard

Downy mildew and powdery mildew thrive in damp conditions at any time of the growing season. Damp, humid autumn days may result in attack of the grapes by the fungus *Botrytis cinerea*, in its unwelcome form of grey rot. Opening up the canopy of leaves and letting the breeze into the vines can help reduce outbreaks. Series of weather stations have been established in many vineyard areas, reporting to a central monitoring service: growers can subscribe to this and have advance warning of the conditions that may result in mildew outbreaks, and are thus able to undertake preventative spraying.

Green harvesting of fruit is generally regarded as good policy; in wet summers when the grapes swell rapidly and the amount of sunshine is reduced, many growers consider it to be essential.

Rain at harvest time is a nightmare, invoking a mood of depression throughout a vineyard area. Grey rot can set in very quickly as the grapes expand by taking up water from the soil, resulting in those in the middle of the bunch being compacted and damaged. Bunch thinning can help reduce this problem. If fruit comes in wet it will be dilute, each grape skin holding water – you have only to shake out a mackintosh after being out in the rain to appreciate how much! Growers at the best of properties have occasionally gone to extraordinary measures to dry out wet grapes – helicopters hovering over the vineyards and giant blow-dryers at the reception in the winery, for example. Mechanical harvesters can prove their worth here, for if rain is forecasted they can get in the crop very quickly. However, many small properties may rely on the use of a contract machine, and this will not be available to them at the time it is most needed. Provided the fruit is healthy, a grower may take a gamble and delay picking until after the rain, hoping for an Indian summer to achieve more ripeness. If hand harvesting takes place, growers may be instructed to leave damaged or rotten fruit on the ground. Sorting tables may be placed at the ends of rows of vines or at the winery reception.

16.3 Handling fruit in the winery

The elimination of unwanted fruit is the first stage of improving quality in the winery. There may be sorting tables at grape reception, or new techniques may be undertaken, such as immersing the grapes in water – the rotten ones fall to the bottom – and then drying the remaining sound fruit. If excess acidity is a problem, the must may be de-acidified using calcium carbonate. As briefly discussed earlier, one recent and very expensive technique is the employment of **must concentrator** machines – these cost up to £80 000 so they are not for the small producer. Vacuum evaporation of excess water results in a concentration of sugars. Machines employing the principle of reverse osmosis, a technique only authorised in the European Union since 1999, concentrate flavours and remove excess alcohol. The largest must concentrator machines can extract 1500 litres of rainwater per hour. It is interesting that the authorities do not consider that this negates the concept of *terroir*, but when two Bordeaux properties covered their vineyards with plastic sheeting during the 2000 growing season, the resulting wines were not accepted as AC (resulting in the world's most expensive Vins de Table).

During the fermentation of red wine, the maker may decide to 'bleed' the vats, by running off some juice, leaving behind a reduced quantity to continue the maceration on the grape skins and concentrate flavours. Oxygenation of the fermenting musts will help soften tannins, which may be very hard in less sunny years, and also reduce 'green' flavours. The length of post-fermentation macerations on the skins will also be reduced in such years in order to balance the tannins with the level of fruit.

16.4 Problems in winemaking

It is obvious that to make top quality wine, the grapes must be fully ripe and disease free. However, during the winemaking process, quality can be lost at any and every stage of the process. We will consider just a few potential causes of quality loss below.

16.4.1 Delay in processing fruit

Grapes should be processed as soon as they are harvested. Ideally, the fruit should be transported in small containers, so it does not crush itself under its own weight. The longer the grapes are left waiting before they are processed, the more risk there is of bacterial spoilage or oxidation. This is particularly important in a hot climate, where the process of deterioration is much faster than in a cool climate.

16.4.2 Lack of fruit selection

Even if the fruit is basically ripe and disease free, invariably there will be some rotten or otherwise poor grapes arriving at the winery. Inspection and elimination of unwanted grapes or materials other than grape is necessary to ensure good wine. In countries where labour is cheap, the cost of staffing a selection table can easily be justified, but in France, for example, many producers of inexpensive wines simply cannot afford such a 'luxury'.

16.4.3 Problems with crushing, destemming or pressing

If the crusher–destemmer or press is old, out of adjustment or not operated carefully, several problems can ensue. For example, it is important that if a destemmer is used the equipment should remove and eject the entire stems. Poor equipment may break up the stems, giving a wine which is very stalky and harsh as a result of the presence of stem tannins. If the rollers of the crusher are out of adjustment, the pressure may be too great, crushing the grape pips and releasing bitter oils into the must.

16.4.4 Lack of control of fermentations

It is important that the temperature is controlled throughout the fermentation period. Winemakers dread a fermentation 'sticking' – stopping before all the sugar has fermented out. This can happen for several reasons: the temperature may have exceeded that at which yeasts will work (approximately 35° to 38 °C), or the juice may be deficient in oxygen or nutrients.

16.4.5 Delays in post-fermentation racking

If, due to delays in racking, the lees start to decompose, then off flavours will be imparted to the wine.

16.4.6 Lack of attention to barrels

Barrels must be clean and sterilised. Sulphur is the traditional means of sterilisation. The belief that wines must be topped up regularly during the first year of maturation is now challenged, but all would accept that oxidation is the great enemy. Fig. 16.1 shows topping up taking place at Château d'Yquem.

Fig. 16.1 Topping up barrels

16.4.7 Poor or over-filtration

Filters must be regularly flushed out, and the filter material replaced as necessary. Poor filtration can lead to cardboard-like flavours in the wine. In the bid to make wines squeaky clean, with no risk of sediment in the bottle which may confuse or annoy the consumer, sadly, many bottlers over-filter the wine. The filter sheets used may be too fine, resulting in a loss of body, structure and longevity in the wine.

16.4.8 Careless bottling

It is important that the bottling process takes place quickly. As vats are emptied, many bottlers blanket the wine with nitrogen or carbon dioxide to prevent any unwanted oxidation at this stage. Although bottles arrive at the plant shrink-wrapped on pallets, they should again be washed before use. 2,4,6-Trichloroanisole (TCA; discussed in Chapter 17) has been found in pallets of supposedly clean, new bottles arriving at bottling lines, which can result in aroma and flavour defects and the rejection of products by consumers.

Common Faults and their Causes

Every day many faulty bottles of wine are opened and, sadly, nearly as many are consumed. Perhaps the drinker is not aware that the wine is faulty, or believes that it is not to his or her taste. Very often there is a strong suspicion that something is wrong, but the buyer does not have the confidence to complain. Detailed below are some of the most common faults encountered in bottled wine. The list is by no means exhaustive.

17.1 2,4,6-Trichloroanisole

'Excuse me, this wine is corked', is a complaint often heard: indeed, to many consumers any faulty wine is 'corked', whatever the actual fault. What is commonly referred to as corkiness is a fault caused by the chemical compound 2,4,6-trichloroanisole. Tribomoanisole (TBA) is a possible cause of similar taints.

This fault can be recognised instantly by nosing the wine – there is usually no need to taste. Upon raising the glass to the nose, the smell is instantly one of a damp cellar, a heavy mustiness, a smell of a wet sack, perhaps with tones of mushrooms or dry rot. If the wine is tasted, it will be dirty, fusty and earthy: rather like biting into a rotten apple. The fault is particularly noticeable at the back of the tongue.

TCA is a chlorine compound and its presence in affected wine is usually, but not always, the result of the bottle cork being contaminated with the substance. Interestingly, wines from bottles sealed with closures other than real cork can also exhibit the fault, thus exposing the myth that contaminated cork is the only cause. Other sources of TCA that may give rise to the contamination of wine include oak barrels, filters, winery equipment and bottling plants. The use of chlorine based disinfectants and even tap water in the winery can lead to TCA taint. TCA is very volatile and migrates easily. Research has shown that cork can act like blotting paper and actually soak up TCA from an infected wine, and thus reduce the level of taint.

There has been much debate about cork taint in recent years, the adversaries being the manufacturers of cork and the manufacturers of alternative bottle stoppers, and the pro- and anti-cork lobbies. Research is now being undertaken by cork manufacturers, trade associations and other interested parties to quantify the number of bottles affected by TCA, and identify its causes. Clearly, the cork manufacturers' aim is to eliminate the sources of contamination of corks.

17.2 Oxidation

Oxidation is a fault that is often apparent on appearance, and certainly detectable on the nose of a wine. A white wine will look 'flat', not at all bright, and in severe cases will deepen considerably in colour and start to look brown. A red wine will also look dull and take on brown tones. On the nose the wines will smell burnt, bitter and have aromas of caramel or, in severe cases, the smell of an Oloroso Sherry. If the wine is tasted, it will be lacking in fruit, bitter, very dirty and short.

At some point in the life of the wine it has absorbed so much oxygen that its structure has been damaged. This may have happened as a result of the grapes having been damaged, or having been subject to delays before processing, or due to careless handling of the wine in the winery. However, the most common cause is when the wine in bottle has been stored badly, or allowed to get too old. Bottles sealed with a natural cork should always be kept lying down to keep the cork moist and expanded in the neck. All wines have a finite life, and should not be allowed to get too old. Inexpensive wines, in particular, are usually made to be drunk almost as soon as they are bottled and do not repay keeping.

17.3 Excessive volatile acidity

The total acidity of a wine is the combination of non-volatile or fixed acids, such as malic and tartaric acids, and the acids that can be separated by steam, the volatile acids. Generally thought of as acetic acid, volatile acid is composed of acetic acid and other acids such as carbonic acid (from carbon dioxide), sulphurous acid (from sulphur dioxide) and, to a lesser extent, acids such as butyric, formic, lactic and tartaric.

As the name suggests, volatile acidity is the wine acid that can be detected on the nose. All other acids are sensed on the palate. All wines contain some volatile acidity. If the level is low it increases the complexity of the wine. If the level is too high, the wine may smell vinegary, or of nail varnish remover. On the palate the wine will exhibit a loss of fruit and be thin and sharp. The finish will be very harsh and acid, perhaps even giving a burning sensation on the back of the mouth.

Acetic acid bacteria, such as those belonging to the genus *Acetobacter*, can multiply rapidly if winery hygiene is poor, thereby risking an increase in the volatile acidity of contaminated wines. The bacteria can develop in wine at any stage of winemaking. The bacteria grow in the presence of air. If a red wine is fermented on the skins of the grapes in an open top vat, the grape skins will be pushed to the top of the vat by the carbon dioxide produced in the fermentation and will present an ideal environment for the growth of acetic acid bacteria. The bacteria can also be harboured in poorly cleaned winery equipment, especially old wooden barrels. Careful use of sulphur dioxide in the winery inhibits the growth of these bacteria, but over use can give another wine fault, as discussed below.

17.4 Excessive sulphur dioxide

Excessive sulphur dioxide can be detected on the nose – a wine may smell of a struck match, or burning coke, that may drown out much of the fruity nose of the product. A prickly sensation will often be felt at the back of the nose or in the throat. The taster may even be induced to sneeze.

Sulphur dioxide is the winemaker's universal antiseptic and antioxidant. As stated above, its careful use inhibits the development of

Table 17.1 EU permitted sulphur dioxide levels

Dry red wine	160
Dry white wine	210
Red wine with 5 g/l sugar or more	210
White wine with 5 g/l sugar or more	260
Certain specified white wines, e.g. Spätlese	300
Certain specified white wines, e.g. Auslese	350
Certain specified white wines, e.g. Barsac	400

acetic acid. However, careless use, particularly before bottling, can lead to the unpleasant effects described above. The amount of sulphur dioxide that can be used in wines sold in European Union member states is strictly regulated by EU wine regulations. Higher levels are permitted for white wines than red, and the permitted level for sweet white wines is greater than that for dry whites. These are detailed in Table 17.1.

17.5 Reductivity

Reductive faults comprise hydrogen sulphide (H_2S) and mercaptans, and are recognised on the nose. Hydrogen sulphide has a pronounced smell of rotten eggs or drains. Mercaptans are even more severe, and the odours are those of sweat, rotten cabbage or garlic.

Reductive defects are caused by careless winemaking. Often sulphur (S) used as a vineyard treatment is reduced to hydrogen sulphide in winemaking by the actions of the yeast. The winemaker should take pains to keep sufficient levels of nitrogen (N_2) and oxygen (O_2) in the fermenting must to counteract reduction. Some winemakers add a minute amount of copper to must, which can prevent the problem, but can lead to problems of copper haze.

17.6 Brettanomyces

Brettanomyces spp. (also known as Brett) are yeasts that resemble *Saccharomyces cerevisiae*, but which are smaller. They are often present in the skins of grapes and cause a wine defect recognisable on the nose as a pronounced smell of cheese, or perhaps 'wet horse'. *Brettanomyces* contamination is a result of careless winemaking and, particularly, poor hygiene management. They are very sensitive to sulphur dioxide and can be inhibited in winemaking by its careful use. Barrels can become impregnated with *Brettanomyces* and thus contaminate wine during maturation. Indeed, whole cellars can be contaminated, a situation that is very difficult to rectify.

17.7 Dekkera

Dekkera is a sporulating form of *Brettanomyces*. It causes defects in wine recognisable on the nose as a pronounced 'burnt sugar' smell, or a 'mousey' off flavour. It is a consequence of poor hygiene in winemaking and can be inhibited by the careful use of sulphur dioxide.

17.8 Geraniol

Geraniol is recognisable on the nose, as the wine smells like geraniums or lemon grass. It is a by-product of potassium sorbate, which is very occasionally used by winemakers as a preservative. It is important to use only the required quantity or the geraniol defect can arise.

17.9 Geosmin

Geosmin is a compound resulting from the metabolism of some moulds, bacteria and blue-green algae. It is recognisable on the nose as the earthy smell of beetroot. Faults which can be attributed to geosmin are normally caused by the contamination of barrels with the micro-organisms that give rise to it, or their growth on cork.

Glossary

Acetobacter. Genus of bacteria of the family Bacteriaceae. The organisms oxidise alcohol to acetic acid.

Acetaldehyde. Product of the oxidation of alcohol, giving a certain aroma to wine. High concentrations give a 'sherry like' smell.

Acid. Essential component – in malic (appley) and tartaric form – of wine, giving freshness and bite. Malic acid is naturally converted to the softer (yogurty) lactic acid via malolactic fermentation, which naturally tends to follow the alcoholic fermentation which has turned sugar into alcohol. Tartaric acid is often added in warm regions.

Alcohol. The product of fermentation, created by yeasts working on sugar. Alcohol is also added in neutral form during or after fermentation to produce liqueur (fortified) wines. Measured as percentage of volume. See Ethanol.

Alcohol by volume. The amount of alcohol expressed as a percentage of the total volume of wine.

Anthocyanin. Grape-skin tannin responsible for colour and flavour in red wines.

Aphid. Small insect which exists by sucking the juices from plants, the most serious of which in viticulture is *Phylloxera vastatrix* (q.v.).

Appellation Contrôlée. Legal status within France's wine grading system (under European Union wine regulations) that guarantees the geographical origin and specifies production criteria for the wine.

Assemblage. French term for the blending of different vats and parcels of wine to make up the final blend.

Autolysis. Interaction between wine and solid yeast matter giving a distinctive flavour, encouraged by ageing the wine on its lees, as in Champagne and some Muscadet.

Balling. See Brix.

Barrel. Curved cylindrical wooden container, historically used for storing and transporting wine. Barrels are made in various sizes (see Barrique), usually from oak, and are now increasingly used for fermentation (of white wines) and/or maturation.

Barrel-ageing. Maturation in barrel in order to add oak, vanilla and toast flavours, and allow a controlled oxygenation.

Barrique. Small oak barrel holding 225 litres (300×75 cl bottles) of wine, traditionally used in Bordeaux; now increasingly used in other regions worldwide.

Bâtonnage. French term for the stirring of the lees in barrels during winemaking.

Baumé. Scale for measuring the concentration of grape sugars, indicating a potential alcohol by volume, commonly used in France.

Bentonite. A type of clay earth for fining.

Blending. The mixing of several wines to create a balanced cuvée. Also called assemblage.

Botrytis cinerea. Fungus that can attack grapes in two ways: in an undesirable way as grey rot, or in a beneficial way as noble rot.

Brix. Scale for measuring concentration of grape sugars, commonly used in the USA.

Cambium. A layer of cells in plants that can produce new cells, including phloem (q.v.).

Cane. Mature shoot of a vine that has overwintered and turned orange-brown. Canes are cut back in winter according to the system of training used.

Canopy. Part of the vine that is above ground, including leaves and fruit.

Cap. The floating skins in a red wine must.

Capsule. Once lead, now foil or plastic-film, which covers and protects the cork and bottle neck. Wax is also sometimes used, particularly for vintage Port.

Carbon dioxide (CO_2). By-product of fermentation. CO_2 is trapped in wine as bubbles by sparkling winemakers using the traditional method, and otherwise induced in, or injected into, all sparkling wines.

Carbonic maceration. Fermentation method using whole bunches of black grapes. Under a blanket of CO_2, enzymic changes within the individual berries take place, resulting in a short fermentation unrelated to yeast. Following this, the conversion of sugars by yeasts continues. Many light bodied, fruity red wines are produced in this way.

Cask. Wooden container (usually cylindrical) used for fermenting or maturing wine and made in varying sizes according to the tradition of the region.
Centrifuge. Machine used to separate wine from the lees after fermentation. May also be used in the production of some low-alcohol wines.
Chaptalisation. Enrichment of must (q.v.) with sugar to increase alcohol levels. The process is named after Dr Jean Antoine Chaptal, a chemist who was Minister of the Interior at the time of Napoleon.
Charmat. Sparkling wine production method in which the second fermentation takes place in tank.
Chlorosis. Vine disorder resulting from iron deficiency in soil, often found in lime-rich soils.
Clone. Group of organisms produced by asexual means from a single parent to which they are genetically identical.
Colloids. Ultramicroscopic particles which do not settle out and are not removed by filtration, and so may cause haze in wine.
Concentrated grape must. Grape juice that has been reduced to 20% of its volume by heating. If 'rectified', it has also had its acidity neutralised. An alternative to using sugar in chaptalisation.
Congeners. The colouring and flavouring matter in wines.
Cordon. Horizontal arm that extends from the 'woody' trunk of the vine.
Coulure. Poor setting of the flowers on a vine, usually due to adverse weather conditions at flowering time (see also Millerandage).
Crushing. Process of breaking open the grapes ready for fermentation.
Cuvaison. Period of time a red wine spends in contact with its skins.
Cuve. Commonly used French term for a vat or tank.
Débourbage. Settling of solids in must or wine either pre- or post-fermentation.
Dégorgement. Removal of yeasty sediment from the bottle at the end of the traditional method of making sparkling wine.
Degree days. System of climatic classification of viticultural areas developed by the University of California at Davis.
Délestage. Translated as 'rack and return', the vat is completely drained of red wine and then the fermenting wine is returned over the cap of skins. It is considered by many to be a gentler process than pumping over, giving softer tannins and deeper colours.
Density of planting. Number of vines planted in a given area of land, usually in Europe expressed as the number of vines per hectare.

Dosage. Following dégorgement, the topping up of traditional method sparkling wine with sweetened wine, according to the intended style.
Egrappoir. Machine which removes stalks from grapes, following which they may be crushed.
Élevage. The 'rearing' or maturing of wines before bottling.
Ethanol. Alcohol in wine, otherwise known as ethyl alcohol (C_2H_5OH).
Fermentation. The conversion of sugars into alcohol through the action of yeasts.
Filtration. Passing the wine through a medium to remove bacteria and solids. Flavours can also be removed.
Fining. Removal of microscopic troublesome matter (colloids) from the wine before bottling. Materials which may be used for fining include bentonite, egg whites and gelatine.
Flavones. Colourless crystalline tricyclic compounds, the derivatives of which are plant pigments.
Fortification. The addition of alcohol to certain wines (e.g. Sherry and Port) either before, during or after fermentation.
Free run juice. Juice naturally drained from solids, from crusher, press or vat.
Green harvesting. Crop thinning before normal harvest. Unripe bunches of grapes are removed to encourage the development and ripening of the remaining crop.
Hybrid. Result of the crossing of two different *Vitis* species, e.g. *Vitis vinifera* and *Vitis labrusca*.
Isinglass. Fining agent derived from fish.
Lees. Sediment, including dead yeast cells, which settles at the bottom of the vat once the fermentation process finishes. Gross lees are coarse sediment and fine lees may settle after initial racking.
Liqueur de tirage. Mixture of sugar and yeast added to base wine at the time of bottling to induce secondary fermentation in traditional method sparkling wine production.
Maceration. The soaking of grape solids in juice. This can take the form of cold soaking pre-fermentation or, in the case of red wines, post-fermentation. Maceration will help extract flavours and, for red wines, colour and tannins.
Malolactic fermentation. A fermentation that may take place after the alcoholic fermentation, in which bacteria convert harsh malic

acid into soft lactic acid. Almost without exception, it is desirable in red wines.

Marc. Skins, stalks and pips left after pressing grapes. May be distilled into Brandy.

Méthode champenoise. Now-defunct term replaced by 'Traditional method'.

Mercaptans. Group of volatile bad-smelling chemical compounds that can occur in wine following fermentation, possibly resulting from over reductive winemaking.

Millerandage. Poor setting of fruit, resulting from poor pollination; consequently the grape berry fails to develop (see also Coulure).

Must. Unfermented grape juice and solids.

Must adjustment. The addition of various substances before fermentation to ensure the desired chemical balance. For example, this may include additions of tartaric acid (acidification), commonly practised in hot climates. Deacidification may be required in cool climates.

Must concentrators. Ingenious machines used to remove water from the juice of grapes. May be used following a wet vintage.

Must enrichment. Process before or in the early stages of fermentation whereby the natural sugar content of the must is increased to raise the alcohol level of the wine. It is common practice in cool climates where grapes struggle to ripen. This is strictly controlled by law in the European Union (see also Chaptalisation).

Must weight. Measurement of richness of must. In practice this indicates the sugar levels contained in crushed grapes or juice (see also Baumé, Brix, Oechsle).

Mutage. The addition of alcohol to stop fermentation – used to make sweet fortified wines.

Oak. Preferred type of wood in which to mature wine. Provides character and imparts flavour.

Oechsle. Scale used for measuring the specific gravity or must weight of grape juice to measure the final alcohol level of wine, commonly used in Germany.

Oenology. The science of winemaking.

Oidium. Fungal disease, commonly known as powdery mildew, affecting vines.

Oxidation. Result of air contact with wine. Controlled in the maturation process; destructive in excess.

Pasteurisation. Destruction of micro-organisms in wine by heating.

Peronospera. Fungal disease, commonly known as downy mildew, affecting vines.

pH. A measure of the hydrogen ion (H^+) concentration in a polar liquid and, hence, the acidity of the liquid. pH is measured on a scale from 0 to 14, where 7 is neutral, below 7 is acidic and above 7 is alkaline.

Phenolics. Group of chemical compounds which includes anthocyanins, tannins and flavour compounds.

Phloem. Tubular structures of specialised plant cells that carry sugars derived from photosynthesis and other products of plant leaf metabolism, from the leaves of vines to the trunks and roots and, importantly, to the grapes where sugars are stored.

Photosynthesis. A biosynthetic process of plants by which carbon dioxide and water are combined to form carbohydrates (starches and sugars). Chlorophyll, the compound that pigments plant leaves, acts as a catalyst in photosynthesis, utilising energy from the sunlight.

Phylloxera vastatrix. Root-burrowing aphid which first devastated Europe's vineyards during the latter part of the nineteenth century.

Pigeage. Punching down of the cap of grape skins formed during fermentation to prevent drying out and encourage the release of colouring matter and tannins. It can be done either manually or mechanically.

Polyphenols. Group of chemical compounds present on grape skins which includes anthocyanins (q.v.) and tannins (q.v.).

Press. Machine used to squeeze out juice from grapes.

Press wine. Wine obtained by pressing the grape skins after maceration. This may be used in blending.

Pressing. Process in winemaking in which the juice is pressed from the skins and other solid matter of the grapes.

Pumping over. The process of pumping the must over the floating cap of skins to obtain more colour and flavour. See also Remontage.

Racking. Transferring juice or wine from one vessel to another, leaving behind any lees or sediment and in so doing clarifying the liquid.

Refractometer. An optical instrument used in the vineyard for measuring must weight of grapes (grape ripeness).

Remontage. Process in red wine making during fermentation when the must is drawn from the bottom of the vat and sprayed onto the floating cap of skins. The purpose is to extract colour and tannins from the skins. Sometimes an aeration is incorporated.

Saignée. A term used to describe the process in which the colour of skins of black grapes has been allowed to 'bleed' for a short period into the juice. Often used to produce rosé wines.

Stelvin®. Brand name for the best-known screw cap closure, now frequently being used, particularly for white wines.

Stuck fermentation. Alcoholic fermentation that has prematurely stopped before the conversion of all sugars.

Sugar. The sugar in fresh grape juice is measured before fermentation as this will determine the alcohol content and style of wine. Scales of sugar measurement include: in France, Baumé; in the New World, Brix; in Germany, Oechsle. In some regions sugar may be added to chaptalise the must – to raise its potential alcoholic strength but not to sweeten it.

Sulphur dioxide (SO_2). Chemical compound used widely in winemaking as a preservative, antiseptic and antioxidant.

Süssreserve. Sweet, unfermented grape juice that may be added prior to bottling to sweeten wines. Often practised in Germany and England.

Tannins. Extracts from red grape skins and oak which give red wine structure. The mouth drying quality of cold black teas is due to tannin.

Tartrates. Crystal deposits of either potassium or calcium tartrate which may form during the maturation period or subsequently. Potassium bitartrate is naturally present in all wine; most is removed before bottling, but some may linger in the form of harmless tartrate crystals.

Terroir. A French term bringing together the notion of soil, microclimate, landscape and environment within a particular vineyard site and the resultant effect on the vines.

Ullage. The air space between the wine and the roof of cask, or in bottle, cork.

Varietal. Wine made from a single grape variety.

Vat. A large vessel which is used for fermentation, storage or ageing of wine. Historically made from wood, but now from other materials including cement, concrete and stainless steel. Some vats are closed, others open topped.

Veraison. The beginning of grape ripening, at which the skin softens and the colour starts to change. Following veraison, the growth in the size of berry is now due to the expansion rather than the reproduction of cells. Sugars increase and acidity levels fall.

Vitis. Grape-bearing genus of the Ampelidaceae family.

Vitis vinifera. European species of the *Vitis* genus from which most of the world's wine is made. There are at least 4000 known varieties, e.g. Chardonnay.

Volatile acidity. Present in all wine, resulting from the oxidation of alcohol to acetic acid (acid in vinegar). Small amounts enhance aroma; excessive amounts cause 'vinegary' smell and taste.
Yeasts. Single cell micro-organisms which produce enzymes that convert sugars into alcohol. They are naturally present on grape skins, or added in cultured form, to ferment the sugars in grape juice.

Bibliography

Allen, M., Bell, S., Rowe, N. and Wall, G. (eds) (2000) *Proceedings ASVO Seminar, Use of Gases in Winemaking*. Australian Society of Viticulture and Oenology, Adelaide.

Allen, M. and Wall, G. (eds) (2002) *Proceedings ASVO Oenology Seminar, Use of Oak Barrels in Winemaking*. Australian Society of Viticulture and Oenology, Adelaide.

Bird, D. (2000) *Understanding Wine Technology*. DBQA Publishing, Newark.

Fielden, C. (2004) *Exploring the World of Wines and Spirits*. WSET, London.

Foulonneau, C. (2002) *Guide Practique de la Vinification*. Dunod, Paris.

Galet, P. (2000) *General Viticulture*. Oenoloplurimédia, Chaintré.

Galet, P. (2000) *Grape Diseases*. Oenoloplurimédia, Chaintré.

Galet, P. (2000) *Grape Varieties and Rootstock Varieties*. Oenoloplurimédia, Chaintré.

Halliday, J. and Johnson, H. (1992) *The Art and Science of Wine*. Mitchell Beazley, London.

Iland, P., Bruer, N., Edwards, G., Weeks, S. and Wilkes, E. (2004) *Chemical Analysis of Grapes and Wine*. Winetitles, Adelaide.

Iland, P., Bruer, N., Ewart, A., Markides, A. and Sitters, S. (2004) *Monitoring the Winemaking Process from Grapes to Wine: Techniques and Concepts*. Winetitles, Adelaide.

ITV (1991) *Protection Raisonnée du Vignoble*. Centre Technique Interprofessionnel de la Vigne et du Vin, Paris.

ITV (1995) *Guide d'Établissement du Vignoble*. Centre Technique Interprofessionnel de la Vigne et du Vin, Paris.

ITV (1998) *Matérials et Installations Vinicoles*. Centre Technique Interprofessionnel de la Vigne et du Vin, Paris.

Magarey, P., MacGregor, A.M., Wachtel, M.F. and Kelly, M.C. (2000) *The Australian and New Zealand Field Guide to Diseases, Pests and Disorders of Grapes*. Winetitles, Adelaide.

Ministry of Agriculture, Fisheries and Food (1997) *Catalogue of Selected Wine Grape Varieties and Clones Cultivated in France*. ENTAV (Établissement National Technique pour l'Amélioration de la Viticulture), INRA (Institut

National de Recherche Agronomique), ENSAM (École Nationale Supérieure Agronomique de Montpellier), ONIVINS (Office National Interprofessionnel des Vins).

Nicholas, P., Magarey, P. and Wachtel, M. (eds) (1994) *Diseases and Pests, Grape Production Series Number 1*. Winetitles, Adelaide.

Official Journal of the European Communities (1999) 14.7.1999L179 Council Regulation (EC) No. 1493/1999 of 17 May 1999 on 'the common organisation of the market in wine'.

Peynaud, E. (1984) *Knowing and Making Wine*. Wiley-Interscience, New York.

Rankine, B. (2004) *Making Good Wine*. Macmillan, Sydney.

Robinson, J. (ed.) (1999) *The Oxford Companion to Wine*. Oxford University Press, Oxford.

Smart, R. and Robinson, M. (1991) *Sunlight into Wine*. Winetitles, Adelaide.

Vigne et Vin (2003) *Guide Pratique, Viticulture Biologique*. Vigne et Vin, Bordeaux.

Useful Websites

Association of Wine Educators (www.wineeducators.com)

The Association of Wine Educators is a professional association whose members are involved in the field of wine education. The website includes a directory of members, many of whom are specialists in different aspects of wine production.

Blackwell Publishing (www.blackwellpublishing.com)

The website of this book's publishers, which includes details of publications and links to numerous resources.

CIVB – Conseil Interprofessionnel du Vin de Bordeaux (professional section) (www.bordeauxprof.com)

The CIVB represents, advises and controls the wine industry of Bordeaux, which is the largest fine wine region in the world. This section of their website (in French) includes 'Les Cahiers Techniques du CIVB' – notebooks containing practical information to assist growers and winemakers.

Decanter (www.decanter.com)

Decanter is a monthly wine magazine, aimed at consumers but widely read by the wine trade and wine producers.

Demeter-International e.V. (www.demeter.net)

Demeter is an ecological association that has a network of certification bodies for biodynamic production.

Grainger, Keith (www.keithgrainger.com)

This website contains contact information for the co-author of this book.

Harpers (www.harpers-wine.com)

Harpers is a weekly journal for the UK wine trade. The website includes précis of recent articles.

The Institute of Masters of Wine (www.masters-of-wine.org)

The qualification of Master of Wine (MW) is regarded as the highest level of achievement in broadly based wine education. The website includes a list of the Institute's members.

ITV – Centre Technique Interprofessionnel de la Vigne et du Vin (www.itvfrance.com)

ITV provides technical information for growers and winemakers in France. The website (in French) includes details of their publications.

Lincoln University Centre for Viticulture and Oenology (www.lincoln.ac.nz)

Lincoln University, based at Christchurch, New Zealand, is a centre of excellence for studies and research in viticulture and oenology. Searching under 'wine' on the website links to courses and staff members.

Office National Interprofessionnel des Vins (ONIVINS) (www.onivins.fr)

ONIVINS is a public organisation that helps develop the rules of vine growing and winemaking in France. This is a comprehensive site, containing statistics and general information, and updates about the French wine industry.

University of California at Davis (wineserver.ucdavis.edu)

The Department of Viticulture and Enology at Davis has been at the forefront of research for over 30 years.

Université Victor Segalen Bordeaux 2, Faculté d'Oenologie (www.oenologie.u-bordeaux2.fr/)

Centre of excellence for studies of oenology in France.

Wine and Spirit Education Trust (www.wset.co.uk)

The Wine and Spirit Education Trust designs and provides wine courses, and is the awards body whose qualifications are recognised by the QCA in the United Kingdom under the National Qualifications Framework (NQF).

The Wine Anorak (www.wineanorak.com)

A lively website containing both objective and subjective views of the wine world.

Wine International (www.wineint.com/)

Wine International is a monthly magazine for wine consumers.

Winetitles (www.winetitles.com.au)

Winetitles is an Australian publisher of wine books, journals and seminar papers.

Index